D1453215

Sightlines

Middlebury Bicentennial Series in Environmental Studies

Christopher McGrory Klyza and Stephen C. Trombulak,
The Story of Vermont: A Natural and Cultural History

Elizabeth H. Thompson and Eric R. Sorenson,
Wetland, Woodland, Wildland: A Guide to the Natural Communites of Vermont

John Elder, editor,
The Return of the Wolf: Reflections on the Future of Wolves in the Northeast

Kevin Dann,
Lewis Creek Lost and Found

Christopher McGrory Klyza, editor,
Wilderness Comes Home: Rewilding the Northeast

Terry Osborne,
Sightlines: The View of a Valley through the Voice of Depression

Middlebury
1800-2000
The Bicentennial
Celebration

SIGHTLINES

*The View of a Valley through
the Voice of Depression*

Terry Osborne

Middlebury College Press

Published by University Press of New England   Hanover and London

Middlebury College Press

Published by University Press of New England, Hanover, NH 03755

© 2001 by Terry Osborne

Printed in the United States of America 5 4 3 2 1

Library of Congress Cataloging-in-Publication Data
Osborne, Terry
    Sightlines : the view of a valley through the voice of depression / by Terry Osborne
        p.   cm. — (Middlebury bicentennial series in environmental studies)
    ISBN 1–58465–083–4
    1. Osborne, Terry—Mental health. 2. Depression, Mental—Patients—Vermont—Biography. I. Title. II. Series.
    RC537 .O79 2001
    616.85′27′0092—dc21   00–012290

*You don't really know who you are until you
know where you are in a physical sense.*
—N. Scott Momaday

BIRDIE

*This book and my love are inseparable,
and both are unchangeably yours.*

Contents

Foreword

Ecologists speak of "edge effect" at the boundaries between diverse eco-systems. Such zones, also referred to as ecotones, can provide remark-ably fertile habitats. From each of the adjacent environments, organisms edge into unfamiliar terrain, crossing the line, as it were, in search of life more abundant. Nature writing is a genre of the reflective essay offering the literary equivalent of an ecotone. Practitioners of the form link an informed appreciation of science and the physical world with an openness to the spiritual, emotional, and aesthetic meaning of nature. Just as it was for Thoreau, personal narrative is often the vehicle of such connection for his successors. They venture out purposefully on their excursions into the unaligned and mysterious woods of home.

Edges widen and proliferate as scientists more closely inspect the subtle gradations within a given landscape. In twentieth-century nature writing, too, the Thoreauvian voice encounters a dramatically expanded range of human situations and concerns. Ranching, gardening, Arctic exploration, urban ecologies, and issues of environmental justice all enter into and enliven this realm of literature. In *Sightlines*, Terry Osborne contributes a powerful and original new aspect to such writing. While exploring the terrain around his family's home in eastern Vermont, he simultaneously begins to identify and map the topography of his life-long depression.

The most memorable and authentic connections in any creative non-fiction are often the unexpected discoveries rather than the insights that first inspired and framed the project—as impressive as those might have been in their own terms. Osborne's book began after he and his wife moved to Thetford, Vermont, he started teaching at Dartmouth, and their two sons were born. His impulse at its inception was closely re-lated to the bioregional philosophy that also lies behind this Middlebury College / University Press of New England Series in Environmental

Studies. It was a sense that, by relating the physical "lineaments" of the land (in Gary Snyder's terminology) with its human history, he could learn to participate more significantly in the life of his family's new home. Such deeper participation coincided, though, in ways that were sometimes baffling or distressing, with a gradual recognition that depression had shaped aspects of his own outlook as surely as glaciation had formed the mountains and carved the rivers of his northern New England landscape.

In writing *Sightlines*, Osborne found that the outer and inner landscapes became one, in ways that were at some times painful, at others exhilarating. His quest for personal integration is convincing for a reader because of its concrete setting in one corner of a particular Vermont town. It is stirring because of the courage, honesty, and humor that suffuse his outings into the surrounding woods and swamp. While avoiding all easy resolutions, he never relinquishes his desire for a healing dialogue with his family's chosen place on earth. I feel grateful for this authentic and beautifully realized book, and believe that many other readers will find themselves unsettled, consoled, and regrounded by it, as well.

John Elder

Acknowledgments

If recognizing my own chronic depression hadn't already shown me how much I needed other people's help, then writing a book about it would have. As I've learned over the last several years, no work like this can be done without the gracious assistance of others.

Deepest thanks to my mother and father for letting me go my own direction and supporting me along a path that must have seemed undirected at times. I learned from them that the best way to love and honor another being is to let it go where it seems to need to go, even if it's not quite clear where that is.

My wife Robin has taught me more things than I can possibly name —among them, faith and love and perseverance—and has shown me above all that a commitment to resolution is at the core of any enduring relationship.

My sons, Carry and Jacob, have graced me with their own sightlines, perspectives far more instructive and interesting than mine. Their presence in this world has enriched my life dramatically, teaching me lessons, particularly about patience and forgiveness, that I never would have learned otherwise. If I've grown up at all in the process of writing this book, it is because of them.

And Griffith, my faithful companion, my trail partner: I can't imagine being in the woods without you.

Ned Perrin—dear friend and mentor—unknowingly rescued my writing life when his own writing ushered me toward nonfiction, where I finally feel at home. His mentoring has taught me creative ways to see the world, and his friendship has been one of the most important in my life.

Audrey McCollum is not mentioned by name in this book, but her influence is everywhere. Her patient and compassionate ear, and corresponding wisdom, pulled me through some very dark years. Without her skill as a therapist, and her understanding as a person, few of the ideas in this book—and none of the introspection—would have been possible.

Ralph Albertini and Tony Dietrich also saw me through some rough times, helped me with their clinical guidance, and managed adeptly my ever-changing medication needs.

In respect to the field work and other kinds of research that have gone into the book over the last twelve years, I'm indebted to more people than I can count. There were the people who let me accompany them out into the field and who shared their knowledge with me. They taught me most of what I know about the land, water, and sky around here: Brian Boland, Charlie Cogbill, David Laing, Ted and Linny Levin, Paul Rezendes, Scott Stokoe, and Tom Wessels.

Many people welcomed me into their offices or homes, or entertained my phone calls, and shared their expertise with me. Some of their names appear in the text, some don't. But all have an important presence here. Among them: Bill Bailey, Ken Bannister, Sergei Bassine, Jeff Corser, Jeff Cueto, Fred Howard, Mark Johnson, Giovanna Peebles, Jeanne Phipps, Dorothy Sears, and Bill Van Fossen.

Marion Fifield, Thetford town historian, is the one who inspired my interest in local history through an adult course I took from her years ago. Charles Latham, Jr., president of the Thetford Historical Society, has written important books on both Thetford history and on Asa Burton, and they have been crucial resources for me. I'd like to thank Marion, Charles, and the rest of the Thetford Historical Society for their support and direction in general, and for the access to the Asa Burton material in particular. I also want to honor the memory of the late Charles Hughes and the late Arthur Bacon; their devotion to Thetford and their knowledge of town geography and history will never be replaced.

Thanks also go to Phil Cronenwett, Robert Jaccaud, Patsy Carter, Marianne Hraibi, and other members of the Dartmouth library system for their willingness to humor me and my crazy needs and then locate whatever book I was looking for.

I'm grateful to all of the students I've been lucky enough to know over the years, both at Lake Forest Country Day School and at Dartmouth College: thank you for everything you've taught me.

And to all of the people who allowed me to wander across their property—people such as Bob and Pat Evans, Bob Watson, Jim Kay, Bill Williamson, the Vaughan families, the Perkinses, the Wakefields—as well as all of those whose property I may have wandered across without them knowing it: thank you for sharing your landscapes with me.

As far as the preparation of the manuscript goes, many people have helped me. Jim Schley of Chelsea Green Publishing was the first to read it, years ago. The transformation it has undergone since then was inspired initially by the responses he gave me. Steve Bodio and the participants of the Wildbranch Workshop, and Richard Nelson and the participants of the Environmental Writing Institute, all provided careful and valuable critiques of portions of this work. Bill Saadeh and John Moody read through specific sections and helped eliminate some inaccuracies in material I was having trouble getting right.

Kate Cohen, first a student and now a friend and valued reader, gave me important feedback on two different drafts of the book. Not only is she a talented writer, she is a gifted editor as well.

I appreciate the amount of time Phil Pochoda of the University Press of New England devoted to the book. He carefully read through it in more draft forms than anyone. April Ossmann, Ellen Wicklum, and the other members of the UPNE staff guided the book expertly through its final stages. Ann Klefstad did a superb copyediting job. And Jim Brisson showed his skill and versatility as an illustrator in the two wonderful maps he drew.

At various times over the last ten years, parts of the book saw the light of day in shorter pieces, and I'd like to recognize the magazines and journals that first published them. The chapter "Uncle Asa" originally appeared in shorter form as "Signs of the Times," in *Vermont Life*. "Mixed Landscape" originally appeared in shorter form in the *Green Mountains Review*. "The Water Drill" originally appeared in shorter form as "The One-and-a-Half-Inch Water Drill" in *Upper Valley* magazine. And "To the Well and Back" originally appeared in shorter form in *The North American Review*.

I've heard other people refer to John Elder—the primary editor of this book—as a "prince," a "saint," an "incredible man." All of those descriptions are entirely accurate and yet wholly inadequate. It is impossible for me to articulate exactly how much John has meant to me over the last two years. He was the first person in a position of editorial authority who understood *Sightlines* as I hoped someone would, who saw the potential in it, and who was willing to put himself on the line for it. Not only does he have a sharp editorial eye, but his manner—compassionate and unfailingly supportive—is one I deeply respect. Suffice it to say that this book's current form, and its publication here and now, are due primarily to him.

Sightlines

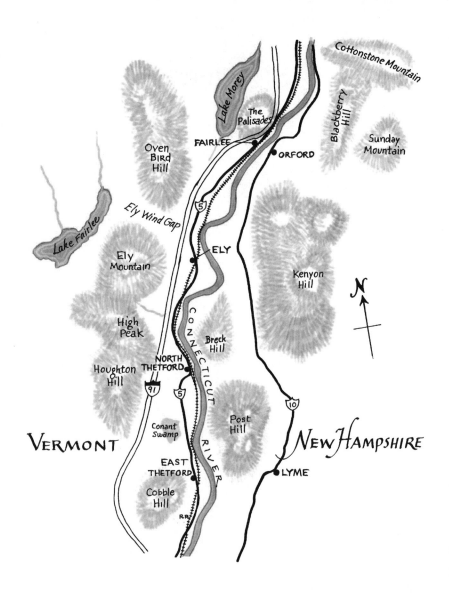

Introduction

The Story's End, November 1999

"Then Frodo kissed Merry and Pippin, and last of all Sam, and went aboard; and the sails were drawn up, and the wind blew, and slowly the ship slipped away down the long grey firth; and the light of the glass of Galadriel that Frodo bore glimmered and was lost...."

I'm stretched out on a couch in my son Carry's room, reading the last page of J. R. R. Tolkien's story *The Return of the King* to him and his younger brother, Jacob, who are lying in their beds. The air is colored with uneven light: behind me a desk lamp gives off enough yellow illumination for me to read; at the far end the blue lava lamp on the table between the boys' beds glows softly; across from me Carry's aquarium glimmers pink.

It's hard to believe we've come to the last page of *The Lord of the Rings* trilogy. Quartet, if you include *The Hobbit*. Four books following the same cluster of characters: Bilbo, Gandalf, Frodo, Sam, Merry, Pippin, Aragorn. Now that we're here I don't really want to say goodbye to them; that seems to be what Sam and Merry and Pippin are feeling too as they watch Bilbo, Frodo, Gandalf, and the others sail away for good.

I continue, "At last the three companions turned away, and never again looking back they rode slowly homewards; and they spoke no word to one another until they came back to the Shire, but each had great comfort in his friends on the long grey road...."

I look over toward Jacob's bed. No movement. He must be asleep. I hope he enjoys drifting off to the sound of a story as much as I love being the one reading it, being the last voice he hears on some nights. Maybe it's self-serving, but on the nights I read to the boys, I feel a protective fatherly connection with them that I haven't felt in many other situations.

When I finish the book, with Merry, Pippin, and Sam having ridden back to their homes in the Shire, where they prepare for a new phase of life, I just lie there for a minute, silent. I'm going to miss them.

Carry peeks up over his blanket. "Is that the end?" he asks.

"Yep," I say. "That's it. Kinda sad."

He lies back down. "Yeah." His voice is muffled under his covers.

"We've been at this series for a long time, haven't we, my man. Maybe a couple of years."

Carry's ten now. We probably started reading different fantasy collections at bedtime four years ago, maybe longer ago than that. The *Narnia Chronicles*, the *Prydain Chronicles*, the *Dark Is Rising* sequence, and now *The Lord of the Rings*: he's been as interested and moved by these stories as I have, and usually just as sorry to have them end. Maybe that's why he seems subdued tonight, though he's not voicing any sadness if he's feeling it. Or maybe he's just on the edge of sleep. Or maybe it's something else altogether and he's just doing what growing boys tend to do —keep more of their thoughts to themselves.

Our ten-year-old. My goodness.

After straightening Jacob's covers and kissing him on the cheek, I lean over Carry and kiss him on the forehead. "Good night, sweet man," I say. Then I give him a squeeze. "I love you."

His arms loop loosely over my shoulders and pull me down a little. "Love you too," he says.

I click off the aquarium light and the desk lamp. The lava lamp will stay on for a while as a night light. Either Robin or I will come back to turn it off before we go to bed.

"Sleep tight," I say as I start to pull the door closed. Then I stop, poke my head back in and whisper, "Only two more days of school 'til Thanksgiving break."

In an English class I was teaching at Dartmouth College in 1988, a year before Carry was born, a Russian student wrote an essay about his return to St. Petersburg after a voluntary two-year absence. Looking down on his homeland through the airplane window, he thought: "My land . . . my country. You know, we cannot choose them."

Those words, and especially their fatalism, jarred me. I understood what they meant—that we can no more pick the place we're born into than we can pick our parents. But I heard them imply more than that. I heard them say that a person's deep identification with *any* place was fated; choice would play no role.

A part of me chafed at that. After all, two years before, my wife, Robin, and I had chosen to move to Vermont from suburban Chicago, where I had grown up. We had chosen a house to rent in the town of Thetford, in the upper Connecticut River valley. And just recently we had moved into another house, one we had chosen to buy. Facing north on a rise called Cobble Hill, the house overlooked a beautiful stretch of the Connecticut River valley; it was in part because of that view that we had bought the place. I felt instantly closer to that landscape than to any I'd known. And so it seemed to me, as I began to explore the twenty-four square miles of valley I could see from home, a landscape I felt inexplicably drawn to, that I had very much chosen where I was.

Now, eleven years later, near the end of 1999, I can only shake my head at my naïveté. I see that the student's words, and all of their implications, were right: there are few things as involuntary as a person's identification with a landscape.

The events in this book took place during a ten-year span, beginning with our move onto Cobble Hill in the fall of 1987 and closing with a hot-air-balloon ride I took over the valley in the spring of 1997. Much of the material has come from journals I began keeping just after we arrived on Cobble Hill. In them I wrote about many different components of my life at the time, but primarily about the walks I took around the valley. I tried to absorb as much of the landscape as I could on those walks, tried to understand how nature worked and figure out what my experiences within it meant. But it was hard going; I'd grown up a suburban boy, and nature was new to me. There was so much I didn't know. So I wound up learning a lot—all of it very basic, and most of it completely unexpected.

I also wrote about our domestic life, about Robin and myself, and our adjustment to a life on our own in a new place, to the responsibilities of a house we owned and a mortgage we owed, and ultimately—with the birth of Carry in 1989, then Jacob four years later—to the revolution of parenthood.

The workings of nature in the valley and the dynamics of life at home inspired a third strand of journal entries that took me completely by surprise. It came from the landscape within myself, and ended up being another exploration—a tortuous trek leading me to a dark but strangely familiar place deep inside. As much as I would have liked to turn back, the trek just kept plowing ahead—as if on its own, propelled by the same necessity that seemed to be urging me out into the valley—and finally stopped beside a chasm at the bottom of me; there, a light was trained on the chronic depression that I suddenly understood had plagued me for most of my life.

If it sounds paradoxical for me to have found so familiar and long-standing a place inside, and still to have been surprised by it, well, it was. But I can't explain it any better than that, except to say that the surprise came not because I hadn't known the place was there; I had just never known it as "depression." Up until then it was simply a voice that had always lived inside my head and come along everywhere with me, engaging me in conversation, inducing me into arguments, overrunning me with bitter commentaries on the world and on my life and behavior.

Now, everyone is buffeted by internal influences of some kind. There are various psychological terms for those influences, depending on the roles they play—an "id," an "observing ego," a "superego," and so on. Some people are aware of them, some people aren't. Some hear them as voices, some don't. But their presence inside us is normal and, in the right measure, even healthy. What distinguished my internal influences from the norm was this: they coalesced into an unrelenting voice that spoke inside my head and sometimes took it over and judged everything harshly and directed those harsh judgments at me and seemed intent on bringing me down and from time to time could actually do it, could get me to believe that what it said about me was true.

I first became aware of all these worlds—new worlds in the valley, at home, inside myself—only after we'd moved to Cobble Hill, and I don't think that was coincidence. Looking back over the years, I can see that the valley had as much to do with it as anyone or anything else. That I'm able to stand here now and reflect on those ten years of my life with some measure of sanity, at least for the moment, is due in large part to the valley's influence, to the way this particular landscape seemed to pull me in and hold me close while the contours of land and water and air imprinted themselves on me, taught me things I'd need to know if I was going to change.

I head downstairs from Carry's room. Our bedroom door is closed, but I can see a strip of light underneath it: Robin's still awake.

"Sweetie?" I say, as I walk in. When I turn the corner, I find her sitting up in bed, reading.

"Hey," she says.

"You're in bed early."

"Yeah, I'm tired. I'm going to get up early tomorrow and get stuff done. The boys asleep?"

"Yep, pretty much."

"You coming to bed?"

"Nah, I'm going to stay up for a little bit."

"You'll turn off the lava lamp then?"

"Yep."

"Thanks," she says.

"Well, then, I guess this is goodnight," I add dramatically and give her a kiss.

"Goooooood night," she says in the deliberately playful way I love, drawing out "good" and dropping her voice down to "night." I've heard her say it that way off and on for twenty-one years now, since we first met in college.

We're in a good place tonight, she and I, feeling calm and able to joke. But that's not always the way things are. There's nothing playful or jokey about the havoc depression wreaks on a relationship or a family, let alone on the depressed person himself. We're all testament to that in this family. In my non-depressed times, like now—and sometimes even in the midst of depressive episodes—that havoc is my deepest regret. It's one thing to experience torment yourself, but it's another, far worse thing to subject your loved ones to it.

If I'm always mindful of, and often appalled by, the hurt I've put them through, I'm also aware of how much I owe them, how lucky I am to be living in their company. I said that my current ability to reflect on a difficult decade of my life is due in large part to the valley's influence. That's true. But what's also true, of course, is that it's due to Carry and Jacob as well, and especially to Robin, whose endurance during these past years has taught me the most important lessons, lessons also inscribed in the valley, though I didn't see them there until just recently: that true resilience is an ability to flex and change; and that true change comes from accepting growth and loss; and that growth and loss have true meaning only when a fibrous love has woven them together. I'm never more aware of that than after a depressive episode, when I see that we've all emerged from the rubble of our emotional lives, bowed and scarred to be sure—different too—but still together, hoping we won't have to go through it again.

From our bedroom I walk back through the kitchen and around the dining room table. Opening the sliding glass door, I step out onto the deck. Over to the right, in the far corner, our plastic deck chairs are stacked against the house in preparation for winter, but there's no threat of winter tonight, so I lift the top chair off the stack and carry it to the middle of the deck.

It's a beautiful, bright night. Earlier, from behind Smart's Mountain, the full moon rose, looking like a swollen orange. It's changed since then: now high in the sky and shrunken to the size I'm used to, it gleams yellow-white, its glow illuminating the Evanses' fields, to our north. Beyond those fields the air is dark, but the valley's out there: Vermont piedmont to the left, the Palisades straight ahead, New Hampshire piedmont to the right. Connecticut River running down the middle. Conant Swamp not far from it. And all of the other places in this mixed landscape I've made my way through—they're out there too. I don't have to see them anymore to know that.

And I don't have to see a thermometer, either, to know how comfortable the air is tonight, especially for this time of year—probably still in the high 40s. It got all the way up to 65 degrees today and it's supposed to be that warm tomorrow too. Thanksgiving week and it's 65 degrees. I believe the global warming theory, but even if I didn't, I'd be hard pressed, during a week like this, to ignore what people around here have been sensing for a generation—that the climate has changed perceptibly.

Still, I empathize with the other point of view—the cautious perspective that isn't prepared to accept or commit to a fundamental change yet—because for several of these last ten years I tried to stake my ground on the cautious side too, at least in regard to my own life. What else was I supposed to do? There I was, going out into the valley, discovering that nature worked much differently than I'd thought; there I was, going inside myself, finding the same thing: that I worked much differently from the way I'd thought. What a shock that was. At first, it was so much easier to take the safe tack and think, "This is just a glitch. Things will get back to the simple, understandable way they were."

But after four or five years it became clear that things weren't going to revert back, because there was no "back" to revert to. What I began to recognize was that the simple pattern of self I'd become accustomed to for the first thirty years of my life had simply been a screen I'd set up to protect myself from the complicated, depressive reality that had actually controlled me.

The recognition of this new reality, the acceptance of change it forced on me, the gradual awareness of its rhythm and texture in my life, and the transformation in thinking and seeing I underwent in the process are all at the root of this book. And they're at the root of the fatalism I now feel about this part of my life, and about this valley. Writer Louise Erdrich describes that feeling eloquently: "We can escape gravity itself," she writes, "and every semblance of geography by moving into sheer space, and yet we cannot abandon our need for reference, identity or our pull to landscapes that mirror our most intense feelings."

I think that's how this story started. When I looked out at the valley the first time, I think I saw, without consciously knowing it, a reflection of my inner self—an undulating image that was very familiar to me, even though I'd never actually faced it before. And because of the "pull" that Erdrich mentions—a geographical affinity that sets us firmly down in places without our knowing why—I felt compelled to stay. Luckily, Robin did too. That's just another part of the story, the fated mystery of our lives. The story is tied together by psyche and family and topography, and the ultimate mystery is the enduring connection that our selves, our loved ones, and the land we inhabit have had all along.

Off to my right, by the butternut tree, I notice a flicker: the tree seems to have gone into shadow. . . . Oh, I know what happened: Rob just turned off her bedside lamp; the tree had been reflecting the lamplight coming through our bedroom windows.

So, the boys are asleep, and Rob's on her way now too. Probably time for me to go. I won't stay out much longer—just another minute. Give myself a little more time to enjoy this bright night and pretend it's the beginning of May. Once I head inside, there'll be plenty to remind me of where I am and what month it really is: there'll be two boys to check on, and a lava lamp to turn off, and doors to lock, and house lights to shut off. Tomorrow, on top of the family routine there'll be students to meet with and winter courses to prepare and holiday plans to confirm. And after ten years, there'll be this book to send to John, my editor.

That's when it hits me: it's been twelve years. Twelve years have gone by since I first stood on this deck and looked at the valley that changed my life. Twelve years. Man, that's a long time. Plenty long enough for this story, for sure. Time for it to be over.

So maybe this is it. Finally. The end of the story. Maybe this is where life turns again, changes direction. Maybe this is where a new story begins. If so, I'm ready.

And when I hear myself think that, I catch an echo of the last thing I read to Carry tonight, Sam Gamgee's voice uttering the final words of *The Return of the King*. He says them after having been away on a long, perilous journey, and after having said good-bye to some of his dearest friends, and after having ridden back to the Shire with Merry and Pippin, the only two companions left from that journey. He returns home to his wife Rose and his baby daughter Elanor, sits in his seat, takes Elanor on his lap, sighs a long sigh, and says: "Well, I'm back."

Part One

Land

Chapter 1

Roots of Air

1

North of our house on Cobble Hill the Connecticut River valley stretches away toward Canada, lined by two states—Vermont to the west, New Hampshire the east—and by two ridges, one on either side of the river. On any map the ridges appear to lie parallel, unconnected, three miles apart. But from our back deck, you'd swear either the map was wrong or your eyes were playing tricks, because the undulating hills aren't straight at all; they seem to sink and taper and curve and finally join each other about eight miles away, resolving this section of valley into an oval bowl —the Vermont piedmont to the left, forming the west rim; the New Hampshire piedmont to the right, forming the east; the blunt, granitic outcrop of the Fairlee Palisades enclosing the far end; and Cobble Hill itself defining the near. Down the middle of the bowl, at its bottom, invisible from our perspective, runs the river itself, the terrain's chisel, once forceful and unpredictable, now dammed to a gentle roll.

The land's contour is what stands out most. It owes its rolling shape to the glacial scrubbing it's had, most recently by a mile-thick ice sheet that pushed south through this bowl about 80,000 years ago and stayed for the next 65,000, before withdrawing back towards the Arctic. During that time the ice depressed the earth's surface hundreds of feet and filed down the tops of the mountains and littered the ground with rocks, gravel, and sand it had carried south with it from what is now Canada.

There isn't the same jaggedness to the ridges here that you find in the Rockies or even in the higher peaks of the White Mountains in New Hampshire. These are just foothills, after all, smaller formations left humped and rounded by the overgoing ice, and so, though they raise the horizon and obscure the land beyond, they do it softly, with a feathered seamlessness between themselves and sky.

The hills are completely forested; the only breaks in the woodlands are the weathered patches on the Palisades' southern face, cliffs too steep and dry to accommodate much but clinging junipers, bristly sedges, and in the warmer months a pair of nesting peregrines. As forest types go, this one could be classified a "transition forest," where the coniferous, boreal forest of more northern latitudes meets the deciduous forest common to areas south of here. In his book *Reading the Forested Landscape* ecologist Tom Wessels describes this region as "a mixing ground where more than one hundred species of woody plants find the limits of either their northern or southern ranges. This mingling of species creates a region with a far greater diversity of plants, and plant communities, than any other area in the northeastern United States."

The forest must have been particularly impressive before Europeans arrived, during the thousands of years the native Abenaki lived in it. We know that its enormous white pines drew the attention of the first French and English wanderers here; so valued were those trees that King George III of England eventually laid claim to them by including in many towns' charters a clause ordering that "all white and other Pine Trees fit for masting Our Royal Navy be carefully preserved for that Use, and none be cut or felled without Our Special License." That Royal order was more or less ignored, and over the next hundred and fifty years the forest, white pines and all, was leveled by European settlers who began moving up to this area from southern New England in the mid-1700s and began transforming the colonial woods into American farmland. During the last hundred years, however, an equally dramatic change has taken place: the farming industry has moved to the wider spaces out West, and so hillside woods around here have regrown where fields and pastures were not long ago. Though today's forest isn't nearly as mature, nor quite as diverse, as the one that stood here three hundred years ago, I imagine it colors and textures the slopes in a similar way.

As wooded as the hillsides are, it would be easy to forget that any open land exists out there. It does, though. Just not very much of it any-more—a fraction of what you'd have seen two generations ago. These days it's confined to the flatter, more fertile ground on the valley bottom, where it's almost entirely invisible to us from our deck—except for a corner of a field we can see, probably three or four miles away, through a gap in the trees—and almost entirely dedicated to cows: fields for hay and corn, pastures for grazing. A handful of families in this section of valley still make at least some of their living from dairy farming, but as I've said, the agricultural way of life that was brought over from Europe and then governed this landscape for two centuries has largely moved West.

Still, for all the predominance of hills in our view, and the absence of open fields, and for all of the shape that river and ice have provided, it isn't finally the land or water that holds sway here, but the sky. The manyfaced sky. It can seem to press over the valley like an airborne ice sheet or fill in like soft meltwater where the ridges crease and sink away, or retreat altogether, leaving the air open and pebbled with the patterns of high clouds or stars. Ours is not a culture that charts its destiny seri-ously by the sun or moon; few of us understand the sky with anywhere near the intimacy that, say, almost every Abenaki did three hundred years ago. So it isn't surprising that I don't know which hill the sun sets behind on any particular day in November, or recognize well ahead of time the telltale cloud progression of an oncoming storm, or notice when a dead star disappears from the night's constellations, or when a new one appears.

But I do respect the sky's presence and influence, the fabric of shadow and light that determines the valley's mood, that often determines mine. I've come to believe during the last ten years that this arrangement of sky over land and water is more than just a stage on which my wife Robin and I play out the scenes of our domestic life; at times it actually seems to direct us. There are days when dark weather fronts pull across to our north, with rainbursts filling the valley like smoke but leaving us untouched, basking in afternoon sun, marveling both at the lucky slant of air that's keeping us bright and dry now and at the twists of life that brought us together as college freshmen. And there are days the sky reverses the divide, when the Palisades' sun-bright slopes against a dis-tant blue seem a cruel tease to us as we labor under a gray overcast, the two of us keeping our distance from each other, feeling as if we've been together too long—I sulking inside a seething depression, Rob tiptoeing around the outburst she knows will come. And there are times when the sky is a single salving cast, as on those nights the emerald swelter of the Northern Lights pulses up from behind the Vermont ridge. Pulling Rob from sleep, I walk her through the house to the back deck, where we stand arm in arm, and beneath the solar breeze and the enormous green ribbons of light fluttering over us like palm leaves, we're inspired to contemplate the mystery of our enduring love.

This valley, then, seems to have built its own character, and influenced ours, through commingled contraries: ground carved by a now unseen river into two straight ridges; two straight ridges bent toward each other to form an enclosure; the enclosure coaxed open by a restless sky; and that sky somehow calmed by the straight curvature of rock. It's a view of freedom gently restrained, gravity on fledgling wings.

Every morning this is the first view I see and every morning it stirs me much the way it did when I first saw it. This valley bowl is one of the few things in life, perhaps the only one, I haven't ever taken for granted, and I can't help thinking that I haven't dulled to it because it won't let me, because it isn't done showing me things I need to know.

Strangely, it wasn't the view that drew me here the first time; it was the house. Ten years ago, in the summer of 1987, I walked casually out of a realtor's office with a secret mission: I was going to sneak a peek at a place I'd just learned was on the market. It took me a while to find it, maybe a couple of days. I suppose I could blame my wandering on the lack of street signs in town then, at least the standard green metal ones with white lettering we have on every corner now; then, there were only wooden ones here and there that people had painted by hand. But the real problem was that I didn't want to ask directions. Part of that was a prideful guy thing: I didn't like asking for help; I wanted to find it myself. Another part was an embarrassment at myself for not knowing where the road was; we'd been living in town for a year, after all, renting a place on Thetford Hill, and I thought I should know. The biggest part, though, arose from a much deeper embarrassment and insecurity: I was afraid that whoever I asked might get suspicious of my nosing around, or worse, might see through me, see me for the self-conscious flatlander I was but was trying desperately not to be.

Eventually, after a day and a half of trial and error—and even subtle direction-asking—I found the road and made my way up it. Rolling slowly into the driveway, a little fearful of running into someone, I rounded the corner and saw the broadsloped Vermont ridge for the first time, risen beyond the house's roofline like a swelling wave. I was awed.

Surprise and luck magnified the awe, I'm sure, because I shouldn't have been rounding this corner at all. Robin and I had already fallen in love with an old farmhouse a mile north of here, a house the owners didn't really want to sell though circumstances had forced them to put it on the market, a place we couldn't afford but that we loved so much we lost sleep fretting over it, imagining whom we could borrow money from and how much of our lives we'd be committing to debt.

Then, by chance, I noticed a listing for this other house in the realtor's office.

"Hey, what's this?" I asked. I was flipping through the listing book absentmindedly, only half paying attention, since I was sure I knew the Thetford listings by heart. But suddenly, unfathomably, a picture of the house appeared. It was as if a leaf, pressed in a favorite childhood book long ago, had fallen from the pages.

The realtor looked up from her work. "Oh, yeah," she said. "That one. I'd forgotten about it."

"Oh, yeah?" a voice inside me was saying. "We've been drivin' all over this goddamn area for the last month, lookin' at houses, and you *forget* about this one right in town? And then all you have to say is, *Oh, yeah?*"

She added, "And the owners are anxious to get rid of it, too."

As she told it, the owners weren't interested in the house at all. They had bought the place for its adjacent land—twenty acres set off to the east that they were going to subdivide into building lots. In the process they had created a separate three-acre lot for the house, hoping they could sell it right away. It was empty and ready to be occupied.

That clinched it. I wanted to see it. I asked for a copy of the listing sheet and said I'd talk to Robin about it when I got the chance. Then, acting only mildly interested though churning excitedly inside, I crept off in search of the house to take an unsupervised look.

* * *

Cobble Hill is a mile-long rise that juts out from the Vermont ridge like a thumb, poking east across the Connecticut valley almost to the river. Its top is a single soft ripple—two crowns separated by a central swale, a graceful undulation between raised ends. Seen from the northeast, the hill seems to stretch out from right to left like a woman lying on her side, the line of her leg leading up to the dome of her hip, then dipping to her waist and rising again along her torso to the higher crown of her shoulders before making a more abrupt scalloped descent toward the river, as if her lower arm were laid out to support her resting head. The house sat just below the crest of her hip.

I followed the driveway thirty yards before coming to a sharp left turn. Blasted through a spine of ledge, the corner was lined on both sides by rock walls—the left overgrown by hemlock, the right by goldenrod. Making that turn, I was blind to everything but the walls.

Around the corner the driveway dropped steeply to the house, so it was just as I came out of the turn that I had the highest vantage, could see everything before starting down. And it was at that moment my heart abandoned that big, beautiful, troubling and unaffordable farmhouse and fell in love with a view of land.

I got out and poked around. The place was very different from the farmhouse we'd been imagining ourselves in. It was a modified saltbox, so its lines were not classically symmetrical, as they were in the other house. It was painted black instead of white. It was twenty years old instead of a hundred and twenty. And when I peeked in through the windows to get a sense of the interior, I could see there wasn't anything near the spaciousness we adored in the farmhouse. I found a narrow galley kitchen, and beyond it a small south-facing sunroom with a wood stove and, connected to them both, a long living room with orange shag carpeting, a built-in bookcase, a picture window and sliding glass door. This place would require a complete reordering of our expectations. Yet I didn't feel disappointed at all; I was already convinced. I was more worried about what Robin would think.

After making a full round of the house, I returned to the back side and stood on the deck, still stunned by the wide spectacle of north-reaching valley. Gazing at it fanned out beyond the neighbors' house and the three black-and-white cows grazing in their pasture, I sensed a deepening of the instant connection I'd felt to this view at the top of the driveway. It was as if I'd come home.

That evening I brought Robin to the house. As we were rounding the corner of the driveway, she gasped suddenly.

"Oh my god!" she said.

"What?" I said, jamming on the brakes, afraid she'd hurt herself or been scared by something. "You okay?"

She was leaning forward, looking out the windshield. "We've got to buy it."

How surprised was I? I thought Rob felt more attached to the farmhouse than I did. I thought I was the only one really dreading the financial burden that place was going to heap on us. All day I'd been readying my sales pitch, thinking of ways to change her mind, to sell her on the advantages of this place and talk up the "positive aspects" of changed expectations. No need for that.

So, how annoyed was I? There she was, the woman who had seemed unchangeably committed to our dream house, looking out the windshield, ready to give it up with her first glance at some other place. "So, tell me," my inner voice was saying to her, "what exactly have we been obsessin' about these last weeks, huh? How important could that house have been to you if you could just toss it away like that? I mean, you are *so* fickle! What's to keep you from changin' your mind about this place? What's to keep you from tossin' *me* away?"

I never said that out loud. Nor should I have, because all that was happening to Rob, of course, was the same thing that had happened to me. But I'd been so afraid she wouldn't like the place or feel drawn to it the way I was, so afraid that I'd have to forget what I'd felt earlier that day, that I couldn't change emotional gears fast enough. The best I could do, sitting in our stopped car at the peak of the driveway, was to feel surprised and righteously annoyed. But the most important thing was that she did like it, she was drawn to it and, like me, was relieved to have found another house to love, one with its own very different, but equal, beauty. One we could afford and actually enjoy.

This is where we live now. This is our home.

What perplexed me then, and does even now, was my instant comfort with the place. It's one thing to find a new view beautiful, and even be stunned by it; it's another to feel strongly aligned with it, to recognize right away a sense of home. Having grown up in Lake Forest, Illinois, a suburb north of Chicago, I knew clearly that this valley was *not* my home. Lake Forest is a residential community built on the wooded plateau above the western shore of Lake Michigan. The terrain of my childhood was lush and gentle, the only abrupt topographical change being a network of deep, narrow ravines that spread from the lakeshore like veins, each ravine tightening and shallowing the further inland it went, and all of them serving as conduits for the land's runoff, draining it all back into the lake.

Around home nature grew at close quarters. Tall, thick-barked oaks and elms shaded our lawn and lined the neighborhood's curving streets. Pines and spruces made fence-like borders between properties. So the only times I saw any distance were when I was riding in a car past the endless cornfields in West Lake Forest or Libertyville, or on my bike three blocks from home, where on the bluff above the lake I might glance east as I was pedaling by, or even stop at one of the cleared lookouts, to catch a glimpse of the horizon, the blended blue of sky and water.

So the view from Cobble Hill did not resemble my childhood home at all. Much more height here, much more distance. And not a single memory. This last quality was something I hadn't considered until I read Scott Russell Sanders's book *Staying Put*. He writes:

> One's native ground is the place where, since before you had words for such knowledge, you have known the smells, the seasons, the birds and beasts, the human voices, the houses, the ways of working, the lay of the land, and the quality of light. It is the landscape you learn before you retreat inside the illusion of your skin.

In my attachment to this valley there was nothing so thorough, no preverbal understanding, none of the once-in-a-lifetime intimacy Sanders mentions. Nor was there anything more mysterious—no haunting aura of a past life waiting to be remembered. No déjà vu. Everything about this place—the breeze, the birdcalls, the light and dark, the rocks, the rivers, the history, the people—was foreign.

Well, not everything. Along with my initial awe at the view there was something recognizable—a broad feeling of familiarity, a calm first-time comfort, as when you meet someone new and know immediately that the two of you will be friends. This is a hard feeling to explain, because it's more intuitive than anything else. It might be the shape of that person's face or the sound of the voice that tells you, or the attitudes or compassion or humor or wisdom. It might be that you're reminded of someone you already know and like, or of someone you wish you were. Or it might be that in this one person are collected pieces of all those cues together.

That's as precise as I can be about this valley. Though I had always felt comfortable in northern New England, the feeling that filled me on the deck that first day had little to do with memory and almost everything to do with anticipation, if that's possible. It was a feeling of longing, I now think: a longing to start again and become someone new, someone more consistent and less moody, someone I could actually like all the time; a longing to feel close to a place by choice instead of assuming it by birth, to understand this beautiful, contradictory landscape— which had stepped unexpectedly out from behind a corner—in a way I hadn't understood anything before.

I didn't know how it was going to happen—at the time, I'm sure I didn't even know that it was going to happen—but one way or another it was going to take some learning.

2

A farmer on the New Hampshire side of the valley told me one day about the trouble he'd had over the years with black bears, how they would come off the piedmont ridge and raid his cornfields. From the pattern of flattened stalks they left behind, he said he knew pretty well what their eating technique was: they'd walk into a field, sit down, reach their arms out and hug the corn into them. Everything within the circumference of their hug was a helping.

This was clearly an annoyance to the farmer, and so I shook my head along with him. But inside I had to smile: there was something so playful and life-affirming in the image of that bear, finding itself in an embarrassment of riches with nothing but time on its hands and a big space in its stomach to fill.

Whenever I think about that bear surrounded by unimaginable bounty, I'm reminded of myself during that first year or so on Cobble Hill, how I reveled in the landscape, learned easily and unworriedly, and pulled life in like a glutton.

Some of the first information came casually, from day-to-day exposure around the house. I began to notice the ordinary left-to-right movement of the weather, and the way the New Hampshire piedmont purpled on fall evenings. I watched the grass in our back field grow four feet tall by midsummer, felt how hard it was to rake up just after it had been mowed. I learned that the black-and-white cows on the other side of the barbed wire fence were called "Holsteins." I saw that the apple trees in different parts of the property bore different kind of apples. I found something that looked to be asparagus flowering like a peacock's crown on the edge of the lawn, far away from where the former vegetable garden was. From the butternut tree out back I felt the sticky softness of the nuts' coat, and tasted the tinny bitterness of the meat.

On my walks, I stayed close to home, traipsing with our dog Griffy through the woods on Cobble Hill, marveling at the various trees, of which I could only identify oaks and maples from the shapes of their leaves. I couldn't tell one evergreen from another. I tested the wetness of the lowlands, stirring up the sulfurous smell of the swamp whenever I sank one of my mud boots in and then pulled it out. Further back in the woods, on a neighbor's land, I bent under lines of blue plastic piping strung from tree to tree, and further still, over another hill, I found a small cabin set against a steep rock outcrop on a south-facing slope. Its door was miraculously open. I crept in and gazed around at the rusted bed frames and tattered chairs, the disconnected wood stove in the corner, the graffitied names of people on the walls and ceiling. I felt as if I'd just stumbled upon a prehistoric cave site.

For the first year or two, I can remember feeling like an awestruck child on those walks. Before I could tell one kind of pine from another, before I learned that the blue tubing was used to funnel sap from sugar maples down to a holding tank or that the cabin was something the neighbors had built as a camp for their kids, I recorded things in the rawest of ways, by sensory imprint and imagination. If there's a purity to experience, particularly as an adult—a way of meeting the world directly, free of the need to name or define things, free of the clutter of thinking too much, free from having to make meaning out of experience—I probably came as close to it on those wanderings as I ever will.

At the time I wasn't aware of myself doing anything intentionally, other than exploring the area around home and enjoying getting to know it. But looking back, I can see that this innocent wandering had the effect of establishing a new "native ground" for myself. It would never be exactly the kind Scott Russell Sanders described, since the valley wasn't my birthplace and I wasn't an infant. But it was my version of it, I guess—an adult version, a reenactment of the original, in which that first year or so of experiences on Cobble Hill imprinted my understanding in a childlike sensory way.

Many walks were full of wonder, and sometimes even a strange magic. On the first day of our first spring here, I snowshoed with Griffy up Cobble Hill from the east end, the "shoulders" end near the river. I wanted to see if I could find my way home across the top of the hill. It's steep on that end, and the snow was thin and uneven, heavy too, so the shoeing was awkward. But I wasn't in any hurry and went slowly, particularly on the upper slope, where the ground was peppered with stumps and branches from recent logging.

Catching sight of a pileated woodpecker as it flew among the remaining trees to hunt for food, I had to keep from laughing and scaring it away as I watched it through binoculars. Cloaked in a long black wing-coat, capped by its red crown, the bird looked unflinchingly stern and ceremonial as it clung motionless to a tree or traveled from one to the next with a shallow swoop and a wing flap or two. But when it moved along a trunk, its aura evaporated: hop hophop hop hophophop it went, straight up the tree, then around it in spirals, making little mincing steps. It was like a Supreme Court justice trying to look venerable while playing hopscotch in his robes.

"Aha," I was thinking. "I see you now. Both sides of you."

At the height of land I headed in the direction of our house, deciding to bushwhack out of the logged area through a border of young trees. The border was narrow, but when I shoed through it, I felt as if I'd walked into the remoteness of a distant ridge. Griffy and I found ourselves in a small square field, lined on all four sides by trees, with a patch of mixed trees and bushes growing like an island in the center. Though tall evergreens threw a dull shadow over the uphill border of the field, the scene was brightened by a smooth floor of snow, its surface an unblemished white. When I looked up I noticed in the distance a familiar view of the valley to the north.

The combined effect was startling. Here was a place on the same slope as our house and no more than half a mile away that faced the same direction upvalley. And yet who could have known it even existed? It was well back from any traveled road and neatly bordered all the way around. I had this childish fantasy of having found a fort where I could secrete myself away, so close to home, yet remote and safe. I could eavesdrop on the world, keeping myself unseen.

I felt something else, though, too, something harder to describe. It was an inkling about the place, a suspicion, like a sharp scent. If I'd known how rarely this pinch of recognition comes, I'd have relished it more. What I sensed was an air of reclusiveness, solitude preferred. A place best left alone. This went against what I could tell about the field from looking around: it abutted a newly logged slope, it was neatly tended by someone, and judging from the barking I heard now and then, it wasn't far from a house. It also went against what I wanted to do, which was to stay and claim the spot, at least to myself. Even so, I couldn't help feeling that the right thing to do was to go. So after keeping quiet a while longer, listening to the emptiness there, I shoed uphill and then halfway across the field, staying as close to the edges as possible, so as not to pock the open snow with prints. Then, turning again, and with Griffy following in my tracks, I shoed off into the woods.

Not every experience was as romantic or certain as that one. Not even close. There was plenty of frustration, countless times when I'd see or hear something and not know what to think, not know what it was. And over time, as I began to recognize how much there was to know and how little I did—trees and birds and rocks, ferns, lichens, insects, all with their own encyclopedic subcategories—the childlike pleasure shrank beneath a growing self-consciousness. In *Nature and Madness* Paul Shepard writes, "The children playing delightedly on the green grass or in awe at an owl in the woods will grow up oblivious to the good in nature if they never go beyond that momentary fascination." So perhaps my deepening dissatisfaction with sensory wonder, and my embarrassment at not knowing the names for what I was sensing, the official terminology for it, was right on cue, the next logical phase of maturation in my new native ground. But it didn't feel that way. It seemed irrational and driven, more self-destructive than mature.

Though this transformation from wonder to antsy insecurity happened slowly, across many walks and in many different places near home, it probably happened most abruptly at the small swamp on Cobble Hill. I went there the first time late one April night, after hearing something through my office window—a steady white noise, like the wash of a distant ocean. Griffy and I followed the sound up the driveway, out onto the road, then up the road a hundred feet or so to the mouth of a grassy corridor leading into the swamp.

The entire swamp was whistling. And it was loud. I couldn't believe how loud it was. Having heard people talk about "peeping," and having recently read a column in our local paper by naturalist Ted Levin, I guessed that I was hearing spring peepers, the inch-long tree frogs that had congregated to mate.

I followed the corridor into the swamp, noticing how the calling in a specific place would stop abruptly as I approached and begin again after I had passed by. Eventually I reached a spot where I was completely surrounded by sound. Though I had a flashlight, I kept it off and amid the whistling relished the volume and pitch. With only my ears as guides, I focused on the sound. I listened to the throbbing of the calls together, a single reverberating mass of noise, high and constant, like the whine of truck tires on the interstate. Then I fixed on individual peeps, listening for the particular cadence of one, and then another, trying to keep them distinct before losing them in the tangle of other calls, this convulsion of male frogs singing for love. It was an overwhelming concert in the dark.

The next time I went back, the calling was as loud and rich and varied as before. But I chose not to notice it; I was impatient to take the next step, to actually see the peepers. For this, however, I was hopelessly unprepared. Ted Levin had subtly warned readers about the time involved in this kind of search: "The longer you squat in the muck, the better your chances of finding a peeper," he'd written. He also recommended shining a flashlight on "rush stems, alder and willow branches, mud hammocks" in order to find a frog—helpful advice, I was sure, except that I didn't know what any of those looked like.

But none of that mattered anyway, because I couldn't even get close to the sound. The same silence that had seemed so interesting on the previous visit wasn't nearly as cute or remarkable this time around. I stepped toward one spot: nothing. Then walked softly to another: more silence.

A third spot: deathly still, while back at the first two places the peepers were going full blast again.

What the hell was going on? Wherever I was, the sound was somewhere else, and it was making me paranoid. I began to feel like a buzzing neon sign, blinking out "STUPID GREENHORN FLATLANDER" to all of those rush stems and alder and willow branches and mud hammocks I hadn't a clue how to identify. And the peepers' loud whistles transformed into catcalls.

The swamp had stirred in me a sharp feeling of incompetence I knew all too well. And hated. And had done my best to avoid over the years. Yes, I'd heard the peepers' kind of mockery before. Lived with it almost every day. It was an outdoor echo of the angry voice I heard inside my head, a voice who spoke to me in words, whose personality was like an unforgiving twin of my own, a Hyde to my Jekyll. He was the one who wanted to lash out at the real estate agent for forgetting about this house on Cobble Hill or criticize Robin for being fickle when she loved the house so immediately. But mostly his attention was turned toward me, his discontent relentless, filled with a searing hatred and derision that made a litany of my own faults, a pronouncement of my every misdeed. And no matter how determinedly I'd turn away or try to ignore him, he'd pull closer and surge and finally fill me like wind to become the only noise I could hear.

"You know somethin', frogman?" he said in his loud, soundless way. "You're nothin' but a stupid shit."

"Is that right," I said.

"You're damn right that's right."

"Well, that's original. Two 'rights' in one sentence. Doesn't that make a wrong?"

"*You* make a wrong."

"Mm-hmm. So what is it now?"

"You know what it is. These frogs, you idiot. How can you not see them? They're all around you! Shit, they're almost jumpin' on you. And you, fancyin' yourself a little Nature Boy. What a joke."

As hard as his angry disdain was to take sometimes, it wasn't unexpected. Over the years the anger had actually become a predictable part of his torment, something solid and consistent that I could brace myself for. What was more dangerous was his funny mockery. He'd start joking and act like a buddy and get me laughing, and sometimes I'd get so involved in the banter that I'd forget what he was up to and lower my guard. Then he'd start in on me for real, and I'd wonder how I could have been so easily fooled. Because the voice never had my peace of mind at heart. The ultimate butt of his every derision, no matter how innocently begun, was me, and the ultimate goal of his every word was to get me to believe him.

More often than not I could deflect his anger or match him joke for joke without feeling too beaten down. But not always. When I couldn't, it was because of his perseverance and power to persuade: to preach and preach and preach until he had convinced me that I was as flawed and hateful as he said I was, or else worse, that I was *nothing*—alone, anonymous, of no importance to anyone, with nowhere to go but six feet under. And on and on and on he would carp, relentlessly, until I was blind to everything but this dark little world of his, this tug of war he'd arranged, the insults he'd rain on me, the arguments I'd have with him—at my desk, in the shower, in the car while stopped at intersections, eventually even in the woods—until it would seem that my only recourse, my best defense, was simply to stop fighting and agree.

The voice had been with me for as long as I could remember. As a kid, I often got down on myself, dwelling on my failures while discounting my successes by attributing them to luck or someone else's charity. That kind of thinking, and even the presence of a damning internal voice, we know today, is pretty characteristic of a depressive mood or personality. Discussing a depressed male patient of his, Terrence Real, psychotherapist and author of *I Don't Want to Talk about It*, writes: "If we were able to take a psychic stethoscope and listen in to the unremitting conversation looping inside [his] mind, we would hear harsh perfectionist judgment matched with bitterness, mistrust and hopelessness."

We may know this today, but thirty years ago the word "depression" hadn't found its way into popular culture, didn't roll off people's tongues as it does now. Like other mental illnesses it was still taboo to talk about, usually confined to whispers beside the words "hospital" or "asylum" and never applied to any but the most troubled children. While that was true in our house as well, my mother could still see that there was something going on with me; she used to tell me that I was my "own worst enemy," not knowing how accurate a description of my inner turmoil that was, and she'd often urge me to make my life easier by not being so hard on myself.

Instinctively, though, I'd already done that. When my own worst enemy became too much to bear, I simply deflected him outward toward other people, a strategy I've since learned is also fairly typical of depressives. As James C. Coyne writes in his introduction to *Essential Papers on Depression*, studies on depressed patients "suggest that their negative feelings, including overt hostility, are also directed at the people around them." On occasion I found not only that the voice could discount classmates to their face as cuttingly as he discounted me, but that when he did I felt better for a while.

I say "on occasion," but the outward-turned voice must have occurred often enough to make it a recognizable part of my personality. I remember the day I discovered that: I was standing next to my parents at home, beside the dining room table where they always spread out the day's mail. They were reading to me from a generally complimentary fourth-grade report card. But under "Weaknesses" it said something like: "Terry could learn to be more tolerant of classmates he feels are not as capable as he." My parents treated the comment supportively, explaining specifically what it meant, what it was to be "tolerant," but that only added to the humiliation I felt at having any identified "Weakness" at all.

That this one-sentence criticism is the only comment I remember from any grade-school report card is indicative of how hard it hit me then and, I suspect, how true I knew it was. It also shows how well the voice has been preserving my failings all these years. But until that day in fourth grade my critical (self-critical) side hadn't seemed so distinctly bad or unusual or weak to me; it was just who I was. From that moment on, though, I knew I had to do something about it. Since it never occurred to me that someone could change himself, even if that self was hurtful to others, I must have figured I was stuck with me. That left only one alternative: to protect myself from the "weak" me. So I did, by convincing myself I had no weaknesses. I created a flawless image of myself, built a thicker shield of affability and self-assurance around it for protection, and then hid deeper inside the battlement, where it was easier to deny the presence of other people's criticisms.

In a very real sense I was two people then. On the outside I wore the disguise of an assertive and gregarious and confident, often *overcon-*

fident, young man; on the inside I was a scared kid chasing success and its half-pleasure as hard as I could, so that I wouldn't have to hear the inner chiding that accompanied any failure.

Success, I learned, could be had most easily on safe ground, by sticking to things I did well. When as a kid it seemed that I had athletic ability, I devoted myself to sports. Football, hockey, baseball, golf: I moved from one to the next with the passage of the seasons, often being elected captain of teams. Of the four it was hockey I had the most promise in, so I concentrated on that. I went to predawn power skating lessons and summer hockey camps and eventually got recruited by and accepted to Princeton, a place I wouldn't have qualified for with grades alone.

And, at the same time, when it seemed to me I had a knack for words and language, I decided to develop that too. At college I majored in English, and then moved back to Chicago to pursue it in graduate school, and to stay close to Robin, who was starting her own graduate work in clinical psychology nearby. A year later I took a job teaching English at the elementary school in Lake Forest I'd gone to myself. Though nothing was ever immune to the voice's criticism, this path was pretty safe: I was back home; I felt at ease in my job and familiar at the school; I understood the students, because I could see myself in them; I knew rules of grammar and syntax; I could talk about stories and arguments, and their beginnings, middles, and ends.

Three years later Robin and I moved east again so that she could finish her clinical graduate work at Dartmouth Medical School. Even though I was moving away from my true home ground, I was excited. We both were. While in college together we'd made road trips throughout northern New England and had always dreamed of living there sometime, if we could find a way to support ourselves.

Once here I made sure to keep things as similar and familiar as possible. I happened into a part-time job teaching English to first-year Dartmouth students and thrived on the challenge of older minds. But a year later we moved out of the house we were renting and bought this place on the rim of a valley bowl, and all my straight-ahead, face-forward, steady-as-she-goes plans got diverted by my attraction to the landscape.

There was a year or so of pure joy, as I've described, of tripping through the woods naïve as a child, when it seemed that I'd found a place where the voice couldn't touch me. But that ended: I must have worn out his grace period for naïveté. He started reminding me that as childlike as I felt, I wasn't one, that I was an adult who knew nothing about the landscape. Nothing.

I suppose I could have turned away from the valley then, retreated back into more familiar territory of teaching and sports, where the footing was surer. But I didn't. I think it was too late by that time: the landscape had already imprinted itself on me. So instead, I tried to quiet the voice by treating the valley like a piece of literature I didn't understand, a skating technique I couldn't do: I decided to learn it, to find names for the things I didn't know.

I started with the most basic information. I learned the place names in the valley, all the towns and villages whose borders fell within the bowl. On the Vermont side, from Cobble Hill north, there was East Thetford, site of the first European settlement in town; North Thetford, which flourished as a rail town after the railroad arrived in 1848; Ely, lying half in Thetford, half in the town of Fairlee, whose train stop, for the last part of the nineteenth century, served the Ely copper mine, nine miles away; and Fairlee, where Samuel Morey, the inventor of the paddlewheel steamboat, spent his final years, never understanding how Robert Fulton had gotten (and taken) credit for Morey's own invention. Heading back south on the New Hampshire side there was Orford, where Morey lived most of his life, and where a line of seven beautiful Federal-style homes, the earliest of which was his, still dominates the Ridge near the river. And south of that, Lyme, one of the most active Underground Railroad stops in the area.

Next I identified the various hills in our view. Following the Vermont piedmont north from Cobble Hill (elevation 873 feet), there was Thetford Hill (939 feet), Houghton Hill (1,472 feet), Potato Hill (or High Peak) (1,703 feet), Ely Mountain (1,468 feet), and an unnamed peak I eventually called Oven Bird Hill (1,569 feet); the Palisades (902 feet) were straight ahead; then there was Cottonstone Mountain (1,760 feet), Blackberry Hill (720 feet), Sunday Mountain (1,800 feet), Kenyon Hill (1,309 feet), Breck Hill (660 feet), and Post Hill (988 feet) coming back on the New Hampshire side.

I bought field guides on birds, trees, and wildflowers. The first wildflower I identified that spring was coltsfoot, a slender, dandelion-like bloom growing beside a path across from our driveway. I learned that one way to tell a Scotch pine from a white pine was by the number of needles in each needle bundle: two for Scotch, five for white. I discovered that the chickadees so numerous at our bird feeder in the winter were black-capped chickadees.

Along with the first bits of bookish knowledge came a renewed sense of accomplishment and conquest. I'd begun learning names for the things I'd been seeing, and given the rules of knowledge I'd grown up under—that nothing was known unless it was named—it confirmed that I finally knew something. In fact, I began deriving more pleasure and reward from identifying what I was seeing than from actually seeing it. And I wanted more of that, and I wanted it faster.

Hurry.

Run from the voice.

Faster.

The quicker I learned something new, I thought, the less likely the voice was to make me feel stupid out there in the valley. And so the speed of the learning became its own measure of success.

Looking back at myself then, I'm reminded of a chipmunk I saw one day in the butternut tree behind the house. Poised on its hind legs, it pulled thick, drooping catkins from the branches, eating them the way we might eat an ear of corn: it held each catkin by the ends and spun it, chewing a single mouth-wide strip around the circumference in quick, choppy bursts, before dropping it and plucking off another. As I watched that assembly-line meal—one catkin, then another, then another—I remember thinking what a waste it seemed, all that hurry, all that catkin left untouched. All of it done with such joylessness.

3

I happened upon the tree near the top of High Peak in the early summer of 1989, on one of my early walks along the Vermont ridge. We'd lived on Cobble Hill almost two years by then, I'd just begun taking my explorations out into the valley, and Robin was within weeks of delivering our first son, Carry.

I'd started my climb on the west side of the hill, the side I couldn't see from home and hadn't explored before, so the ground seemed charged, potent with unknowns. And it didn't disappoint. It surprised me with its topography, the spine-and-gully formation lying across it like ribs that took me up, then down a little, up further, then down a little again, like a tide carrying me out from shore. It surprised me with two peaks at the top instead of one, paired knobs separated by a hollow. And it surprised me with its high openness, made by recent logging, which had cut a swath up the south side of the hill into the hollow.

Later, after coming off the peaks, I wandered along the west edge of the ridgetop for a while. Eventually, I found myself in a gully with a raised wall on my left that sloped uphill away from me. Concentrating on my footing, I had my head down most of the time, but at one point I glanced up to find the most surprising sight of all: a fifty-foot tree in full leaf, its base raised five feet off the ground. And below it, cast in patchy sunlight, where its roots should have been, was the leaf-lined trough of the gully, and nothing else. In the instant it takes to be startled, it occurred to me the tree had no roots, or else roots of air.

After the initial shock, I walked closer and investigated, trying to understand what had happened. It looked as if the tree, a maple, had been blown down years before, almost entirely uprooting itself in the process. Its top had come to rest on the uphill ground to my left; down by me, it had fallen across the gully like a drawbridge across an empty moat. A branch that had been perhaps ten feet off the ground when the tree was standing suddenly found itself upright above the gully when the tree fell over, and it just kept growing. Now, years later, it had become its own tree, taller and thicker than the fallen one out of which it had grown and which supported it now, serving as its base, a cylindrical platform five feet above the gully floor. I raised both my arms onto that fallen trunk, and leaned against it.

"Pretty little theory," the voice said.

"You like it, huh," I said.

"Oh, yeah. A branch growin' into a tree on top of another tree, and all without roots. That's priceless."

"Well, that's what it looks like, doesn't it?"

"Not if you know anythin' about trees."

"So you know about trees?"

"One of us has to. Listen, it's simple: if you've got a live tree, you've got roots."

"I don't even know enough about trees to say whether that's true or not. Aren't there any kinds of trees that stay alive some other way?" I asked him.

"Oh, please," he said.

"Okay, then where are this tree's roots? They aren't underneath it."

"Man oh man. You're not really that thick, are you?" he said, then added slowly, in a singsongy way, as if to a young child, "Why don't we try lookin' where the rest of the roots are?"

The original tree's roots were to my right, but from where I stood the entire root base looked to have been upturned, yanked out of the ground long ago when the tree had fallen. Still, some of those roots could have held, I guessed, at least enough of them to feed the new tree. But when I walked around the base, I found the underside completely exposed—a giant earthen hand, upright, looming taller than me, its torn roots poking like spindly fingers from the edges of its veined palm. The more I surveyed the broad underside of the base, the less I could explain how the tree was still living. There seemed too much rootedness open to the air, too little still planted to support both the new tree and the solid platform of the original one.

"Here we are," I said. "Now, show them to me."

"It's not the kind of thing you can just show. You dig underneath the base, you'll find 'em. Go ahead. I guarantee you they're there."

I said, "Guarantee, hell. You're not really sure, are you? You could be wrong."

"Not likely."

"But possible. For all of the guaranteed roots you've shown me, they might as *well* be growing in the air."

I meant it, too. In the shadow of that base, with no live roots visible and all my other explanations shot to hell, I half-dared to believe that magical airborne roots were possible.

And, in fact, they are. In *Volcano*, Garrett Hongo's memoir of his return to his childhood home on the island of Hawai'i, he describes the trunk of an enormous native tree fern called *hāpu'u*:

> Its stump is its root, a bundle of matlike fibers that take on sustenance from the rains. The standing trunk is actually a bundle of fibrous stalks

. . . surrounded by an absorbent, spongy matting that knits everything together. This matting is, in actuality, a system of intertwined aerial roots. . . . The villagers here in Volcano know that you must water *hāpu'u* from the *top* of its trunk, not at its base.

Later, a friend tells Hongo about *hāpu'u*'s life cycle:

The *hāpu'u* gets to a certain point when it's too old or too heavy to support its crown—about thirty or forty years. Then it collapses, and you'd think it'd die, but it just sends up another shoot and makes a joint back from the fallen trunk up to the canopy again and lives for another thirty or forty years until it collapses another time and sends up a new bundle of roots and fronds. A tree can fall a bunch of times, make a series of joints like you see around here and send up a new crown every time. . . . It's *immortal*!

I didn't know about *hāpu'u* that afternoon on High Peak. I would have wanted to, I'm sure, but it's good that I didn't. I probably would have used it as I used other book-fed information in those days—to squelch my insecurity and answer the voice. Instead, without a concrete explanation for the floating maple tree, I had to rely on the thing that had first propelled me so happily through Cobble Hill's woods: raw wonder. And, god, how freeing it was then to remember that uncertainty didn't have to turn savagely inward, that it was still possible to revel in something's impression, and *believe* it, even if for an instant, no matter how fantastic the impression might be; I felt like a child again, out ahead on a walk, conjuring the ordinary into the magical before the doubting adult arrived.

So when I learned of it, I greeted the existence of *hāpu'u* with the wonder it deserved, and not greedy affirmation. What a miraculous world it seemed, to have room for a tree like that, growing upside down and rightside up at the same time, both its roots and leaves in the air, rising from its fallen self like a phoenix from its ashes.

I also greeted it with gratitude, for if something like *hāpu'u* could exist, then so could a floating maple, and so too a New England valley bowl where straight ridgelines managed to bend themselves and meet. If *hāpu'u* could flourish, then perhaps there was hope for a man who had pulled himself from his home in the suburbs of Chicago and, chased by the enemy voice inside him, suddenly felt like an insecure boy in that valley. Perhaps he could discover home there and grow securely into it. Perhaps he could learn from its contours the complex ways of the land, and then those ways could teach him how to soften and bend the lines of his own divided nature—lines that made him feel split in two sometimes, bouncing back and forth between boy and man, dumb jock and capable teacher, greenhorn flatlander and valley nature lover, hurried chipmunk and unworried bear. Perhaps he could come to understand through his own experience what he had dared to believe for a moment one afternoon on a hilltop: that a living thing could find fertile ground where you might least expect it, in the invisible firmness of air.

Chapter 2

Uncle Asa

I

When we moved into our house in the fall of 1987, the place across
the road was one of the oldest farms in Thetford. Two months later it
was sold to a developer. There was no warning or anything. We couldn't
believe it had happened.

Tended primarily by four families over two hundred years—the Bur-
tons, Slafters, Chases, and Chiotts—it had grown into a beautiful mixture of land. There were a hundred and twenty-five acres in all: thirty
uphill acres of pine knolls and swamp; below them, twenty-five open-
field acres sloping down to the farmhouse and barns; and on the west
edge, along the interstate, a seventy-five-acre strip of mixed woods designated as deer yard.

The farmhouse itself was the oldest one left in town. It was started
in the summer of 1779 for Asa Burton, a twenty-six-year-old graduate
of Dartmouth College who'd had only three months of formal divinity
study and a year of practical ministry under his belt before being invited
to become the pastor of Thetford—the second pastor—succeeding
Clement Sumner, a Yale graduate who was forced out of town because
of his Tory loyalty to England. After some hesitation Burton accepted
the invitation, and so in January 1779, as he wrote in his autobiographical papers, "I was accordingly ordained over the Church & Society of
Thetford."

As part of the ordination package, the residents of the town agreed to give Burton an initial £50 payment (which went toward building the house), a fifty-acre lot, twenty-five cords of firewood per year, and a salary of £85, or $283.33, each year. He lived in the same house on the same piece of land and on no more than that original salary for most of the next fifty-seven years, until his death in 1836.

Burton was a small, intense man described by one writer as having "a retreating forehead, a bright radiant eye, and a nose that covered about one half of his face." A stern Calvinist, he believed in the inherent sinfulness of all humans. But he also believed that each of us was salvageable, as long as we underwent a conversion, a moral rebirth, by maintaining a perseverant, single-minded faith. For him a holy life was one spent serving God, obeying His laws, and abstaining from earthly desires. Salvation was no guarantee, since God alone knew the rightful members of His elect, but in the face of this uncertainty there was a reliable way to measure a person's character and his or her fitness for election: outward behavior. As he wrote in one of his published essays, echoing Christ's own words,

> The heart is every person's moral nature; and his external conduct is the fruit this nature brings forth. . . . This is the way by which we learn what the nature or moral character of man is. We infer his nature from his fruit, as we judge of the nature of a tree by its fruit.

In other words, external behavior revealed internal nature.

When he arrived in Thetford, Burton found the residents unabashedly sinful, inside and out. "The inhabitants in general were rather poor, vulgar in their manners, far from neatness in their appearance within doors or without, & their lives, a few excepted, were immoral & disgusting," he wrote. "Many heads of families were intemperate, hard drinkers of ardent Spirits. They were very fond of amusements; especially such as visiting each other for the sake of vain & merry conversation, & drinking. They had very little or no family government. . . . It seemed to me the gospel could never have any success among them, till they were more reformed and civilised."

The Thetford landscape in 1779 was not much more civilized than its inhabitants and just as in need of reformation. That shouldn't be surprising, since Thetford had been a town for only eighteen years. It was in

1761 that Benning Wentworth, governor of New Hampshire, had established the town's charter and granted it to sixty-two "proprietors" (they
were land speculators, really, most of whom lived in Hebron, Connecticut). Three years later, in 1764, John Chamberlain—an agent for one of
those original proprietors—headed north from Connecticut and picked
a spot in the southeast corner of town to live.

By the time of Burton's ordination, there were about four hundred
people living in Thetford; only two families, by his count, had settled in
the hillier, western half of town. Three mills had been established as
well: at one site on Gun Brook in North Thetford there was both a
sawmill and gristmill, while on Zebedee Brook in East Thetford, near
John Chamberlain's original lot, another sawmill was operating.

The mills' existence underscores the work of settlement taking place
throughout the region then: trees were being cut, newly opened land
was being cultivated. But clearing the forest by hand was no treat. Burton knew this from his own experience, having spent much of his early
life doing just that at his father's home in Norwich, the town just south
of Thetford. "At Norwich I lived with my father & spent my time clearing land. . . ." he wrote. "During this time I labored hard, & brought
with other help about sixty acres of wild land to a state of improvement.
During this time from 14 till I was 20 years of age, my principal labor
was chopping timber, rolling it into heaps to burn, & making log fence.
The work was very heavy & wearing to the constitution. I several times
strained my breast lifting to such a degree, that at 19 years of age, I was
not able to labor but little compared with what I had done."

So his spirits must have dropped when he saw the land set aside for
him in Thetford: "The land given me was wild; not one foot cleared on
it when deeded to me."

As for the house the town built for him—well, it was no Monticello.
Originally the residents "put up a frame, covered the body, & inclosed
the south room with rough boards, & shingled a part of the roof, dug a
well, & Cellar, & stoned them, & built a chimney. . . . There was only the
south room habitable; & this had but one small window in it. In this situation I lived nearly two years."

Through his hard work over the first years Burton demonstrated in
practice what he believed in principle—that humans, though fundamentally depraved, could improve and become more open to God's

grace. By preaching pointedly and consistently against all sinfulness, by ousting members of the church whose conduct did not live up to God's standards, and by holding special meetings to teach the members about religious music and singing, he gradually molded the people into a flock with whom even he was modestly pleased. "The inhabitants became more civil, polished in their manners, moral, & regular in their conduct," he wrote. "So that after a few years they were considered abroad as regular a people in all respects as any on Connecticut River."

He found the land itself just as amenable to improvement: "The land was fertile, & as soon as cleared & sown or planted, it yielded from 20 to 30 bushels of wheat or Rye to the acre, and from 40 to 50 bushels of corn. . . . Also where the land was sown with clover, or herdsgrass seed it was not uncommon for it to yield three tons to the acre, & some land even four tons." With the help of some townspeople, he began clearing his own land, an acre or two per year. Then he planted wheat, rye, and grass seed, until, as he wrote, "in two years [I] was able to raise grain sufficient for myself, and cut hay sufficient for a Cow & horse."

Improvements to the house, however, came less dramatically. Two years after it was built, nothing new had been done, so he took the initiative himself: "I was obliged on foot to go to Lyme, where their meeting house now is, & purchase nails, & bring them home on my back in a keg I had for the purpose; & then with my own hands shingle the remainder of the roof, all the back part, & one half of the fore part of the house. And that shingling has remained good to this day, above forty years."

2

It was two centuries after that shingling, in the months after the farm was sold to the developer, that I began my first wonder-filled walks around Cobble Hill. All of them wound through Asa Burton's grounds. By following already worn trails over the high-ground woods to the south and through the swampy terrain to the east, I established a loop I walked with Griffy almost every day for a year. And after getting used to the sights on the loop—the enormous maple spreading over a height of land, the tall shadowy conifers beyond it, the blue plastic piping strung like a spider's web uphill from them—I made side trips, little branching explorations into the more remote reaches of the place. On foot in the fall I came across the shell of an old truck back in the southwest corner of the property, abandoned a long way from anything a long time ago; trees were growing through its glassless windows. On skis in the winter I kicked through shin-deep snow in the deer yard, wandering through hemlock shadows, listening to the maples clacking together in the wind, following here and there the dotted tracks of deer and rabbit. By our second spring on Cobble Hill, I'd started to feel close to those woods.

But all the while changes were being made. They began benignly enough the first spring, when a surveyor came and staked out the six individual lots for the development, marking the property lines with bright orange flagging. There was a pause until summer, when an excavation crew arrived and blasted out a driveway right across from ours and then dug out a fire pond on the upper edge of the field. Finally, the following winter loggers rolled in and selectively logged trees out of parts of the deeryard.

That all seemed bad enough, but early the second spring, the spring of 1989—just as I'd begun feeling at home in Burton's woods, just after I'd found the floating maple on High Peak, and a month before Carry was born—the developers outdid themselves: they placed two signs at the mouth of the development's driveway. One was a green metal street sign with white lettering, stuck to an eight-foot green metal stake. It read "Long View Drive"—a remarkably creative name, the voice and I agreed, one the developers must have spent at least twenty seconds coming up with. The other was a large square maple sign with an ornamental forest-green border and thin green lettering, spelling out the name "Burton Meadows."

I didn't realize how angry I was about having that land transformed into a development until I saw the wooden sign. It was a hokey resort-style sign in a nonresort town. It had no directional value, hidden as it was off to the side of a dirt road a quarter-mile from the main road. (Although, I did give the developers some grudging credit for the placement. They seemed to understand that down on Route 113, where it would have been more practical, their sign would also have been a juicy target for the drive-by vandals that feasted on our mailboxes.) And it was an exaggerated, costumed bow to local history, like some supermarket calling itself "Ye Olde Towne Shoppe." As if somehow the development was more acceptable, more legitimate, because it included Asa Burton's name.

"That thing is hideous, " the voice said.

"Worse than that," I said. "It's . . . quaint."

"Anything that ugly's gotta come down."

"Yes, it does. And in a hurry."

I wanted to take a chainsaw to it and level it right away, a testament to my new chainsawing ability and a mini-monkeywrenching act in the spirit of writer Edward Abbey, for whom sabotage was a legitimate response to development: "Oppose," he once wrote. "And if opposition is not enough, resist. . . . And if resistance is not enough, then subvert." That advice carried the emotion I felt, but finally I had to admit that my personality wasn't quite so daring. So I decided against the chainsaw. It seemed a little knee-jerkish anyway, too obvious for the neighbor across the street to do, and a little too much like adolescent vandalism for my own conscience.

Still, I had to do something. And finally, an idea came. At the mouth of our driveway I planted my own kind of resistance. I took the rattiest piece of plywood I could find, grayed from dirt and age, nailed it to a pair of weathered two-by-fours, and used leftover barn paint to write, "Uncle Asa's Nuclear Waste Dump and Tick Farm: Keep Out!" After I sketched a scraggly red border, I asked Rob—eight months pregnant—to put the final touches on it. She added a skull and crossbones near the bottom.

When she finished, we both stood back and looked at it. Man, I loved that sign. It was ugly. It was weird. It capitalized on a couple of local concerns at the time: nuclear waste disposal and Lyme disease. It perverted Asa Burton's legacy just as badly as the "Burton Meadows" sign had. And if you knew Robin and me, how unlike us it was, it was pretty funny too.

"Beautiful," I said to her, looping an arm around her shoulders. "A masterpiece."

"Thanks," she said and returned the hug. "Now, you're not really going to put it up, are you?"

"Yes, I am."

"Isn't it illegal or something?"

"I don't know. It's going to be on our property."

Shaking her head slowly, she said, "I can't believe it. We're about to be parents and we're going to do something like this."

"What will our child think!" I said in mock horror.

"It'll think the same thing I think."

"Which is?"

"Which is that I don't know how the sign got up there, and for that matter I have no idea who you are either."

"Complete disavowal of any criminal wrongdoing."

"Exactly," she said, then smiled sweetly. "Okay?"

"We never met," I agreed, then looked down at her belly. "Well, I guess we met once."

"Not that I remember."

At first, the sign was simply a way for me to vent my anger and show my opposition to the development. In that respect it was a success, if only on a small scale. Other people living on the hill loved it. Some even instructed their friends to drive by and see it. One propane delivery man told me that on his first trip to our house, he stopped outside the driveway and radioed his dispatcher to ask if he was at the right place and if it was safe to go in.

The immediate attention was gratifying, I admit, but secretly I imagined even bigger things. Secretly I dreamed the sign might have some wider influence and really hinder the development—maybe the developers would remove their signs, or people would think twice about buying a lot in Burton Meadows if it meant living across from someone who stored nuclear waste and raised disease-carrying ticks, or who was sick enough to joke about it. And surprisingly, no lots did sell. We didn't know why, but without any better explanations we gave the credit to Uncle Asa.

But enlisting Asa Burton's name in support of my protest was ironic at best, because he would have firmly disagreed with my position. Not with the protest itself, necessarily. Resistance and opposition had a place in his world view: against enemies of Christianity he counseled hatred and advocated their destruction. What he never would have supported was my desire to save the woods. Wilderness was uncivilized, disordered, and as a result, immoral; it was antagonistic to the light of Christianity. As far as he was concerned, the less of it the better.

I don't think he ever really looked at the land, even his own land, for its intrinsic qualities. In his writings he referred to the landscape primarily for its moral metaphors. Whenever he mentioned the details of his farm, it was to convey a lesson about his prudence, frugality, and care — the way each tool had a designated place in his barn, and the way each was kept in good repair, thereby extending its life and avoiding the need for new tools.

Only twice did he ever mention the land itself, both times in the context of a daughter's death and his own consequent grief. But even in those cases there was a moral lesson behind it. Describing the fatal illness of his oldest daughter Lucena at age six, Burton recalled a wildfire burning across some of his property at the time:

> I recollect viewing one day in my house, seeing how it raged; that I then thought, if my daughter was not sick, & we all enjoyed health, I should view that fire a great calamity, & should be much concerned about the damage it might produce. But compared with her sickness & death, if she should die, as I then expected; that calamity appeared to me as nothing; & it in fact gave me very little concern, or trouble; tho many hands were fighting it, & it was constantly destroying my property, as well as the neighbors'. But the loss of worldly property compared with the loss of my child appeared as nothing.

Eleven years later, his second daughter Polly died, leaving him then childless. (A third daughter, Mercy, would be born five years later.) Looking back on her death, he tried to describe his grief in some concrete way, to put his "feelings after her death in relation to worldly property." He recalled one particular event, telling it almost as a parable:

> A neighbor had some land adjoining mine, which he knew would well accommodate me, & which I had wished to purchase; but he would not sell it. Some time after my daughter's death he came to me one day, & offered to sell me that land, & urged me to purchase it. I at last made this reply; Sir, I feel no wish now to purchase it; indeed, if the whole Town of Thetford were offered to me for sixpence, I would not be at the trouble of buying it.

Bowed by grief after his daughters' deaths, Burton used the land to put life in perspective. For land was, to Asa Burton, quintessential worldly property, something never to be overvalued for itself. What was important about it was the way it could be used for moral purposes—the way, according to God's dictum, it could be subdued and have dominion held over it, and then be incorporated into an already moral and upstanding life. A person who managed his land properly was performing God's will; righteousness would emanate from that land like a holy sign, a glow visible to everyone, a glow that would affirm the quality of that person's character.

3

With Uncle Asa at the mouth of our driveway, nothing happened in the development for the next two years. The signs simply stood there facing each other placidly across the road. That was probably good, because things at home were anything but quiet. Carry's birth had thrown Robin and me for a loop.

We thought we were ready. Rob had studied all the recommended childrearing books, reading excerpts to me at night in bed. We'd talked a lot about our roles, about how each of us would juggle work and parenting. We'd watched our friends handle their first children and listened to their experiences. But to say we were unprepared for the revolution of parenthood would be an understatement.

The hardest part for us was Carry's initial unhappiness. After a long, hard labor he emerged from Robin without a sound, his face pale blue, his lungs still: he wasn't breathing. The medical team outside the delivery room revived him quickly, but once they did he sounded none too pleased about the whole experience. That displeasure lasted for months. Whether for his daytime naps or nighttime sleep, he'd go down crying and wake up that way too, and from about five to nine each evening, a period we started calling "the hours from hell," he'd be colicky and inconsolable.

We tried anything to comfort him. Over time we each found techniques that were moderately successful. Mine was an elaborate assortment of moves I called "Boober Doobers": I'd lay him face down across my hands, stretch my arms out in front of me, and then do deep-knee bends, swooping my arms a little with each bend. Usually, seventy-five Boober Doobers would quiet him, and finish me.

Carry's discomfort, we now know, was due at least in part to the sensitivity he inherited in double measure from Rob and me—a tender sensory awareness from Rob, and a raw emotional sensitivity from me. Every stimulus affected him so acutely in those first months that it must have felt to him as if he were without skin, or covered with only a thin film of fluid. This, as I say, we know now. At the time we didn't, and it left us wondering whether his biggest challenge in life wouldn't be overcoming his parents' cluelessness.

Robin was much better at nurturing Carry than I was: she was more patient, empathic. But because of that, because she took on each of his discomforts as her own, she was emotionally exhausted at the end of every day. So was I, but for a different reason: I felt a failure. Before Carry was born, I was sure I'd be the ideal dad—a compassionate, stay-at-home man who could handle placidly anything his child threw at him. Two months later, I was a wreck, my image of my fathering self shattered. Oh, there were magical times when I'd rock Carry to sleep after he'd eaten and then keep rocking him gently in my arms, listening to his breathing quicken, watching his expressions change with his dreams— his eyelids quivering, his brows furrowing, his mouth turning up at the corners, sometimes a little sucked-in giggle sounding in his throat— wondering how an infant could make faces in sleep he hadn't learned to make while awake. But those moments were quickly forgotten. What stuck in my head were the times I couldn't calm him, so many times, when I'd try and try and try but still couldn't do it. Particularly haunting were the handful of times I'd lost my patience and raised my voice, held Carry out at arm's length and yelled at him to stop crying, *yelled* at this infant who wanted nothing more than to feel at home in the world.

If the peeping tree frogs in the swamp had made me feel insecure and incompetent, if the voice's constant mockery could wear me down, Carry's cries those first months felt exponentially worse than either: not only could I not just walk away from them as I could from the frogs, but I couldn't relegate them to background noise either, as I sometimes could with the voice. Carry was my child, my blood, my responsibility. He was depending on me, and there I was getting mad at him when I couldn't meet his needs. *Because* I couldn't meet his needs. What a pathetic idiot I thought I was.

In his paper "Mourning and Melancholia," Sigmund Freud pointed out some distinguishing characteristics of depressed patients (or "melancholiacs," as he termed them). One of the most pronounced, he noted, was a fall in self-esteem. A depressed person "represents his ego to us as worthless, incapable of any effort, and morally despicable; he reproaches himself, vilifies himself, and expects to be cast out and chastised. He abases himself before everyone and commiserates with his own relatives for being connected with someone so unworthy." Freud might as well have been describing me and my state of mind in the months after Carry's birth. Of course, I didn't know that I was thinking like a melancholiac, but even if I had known, I would have denied it. There was still very little room inside me for weaknesses, and none for a diagnosable one like "depression."

All I knew was that fatherhood was like having a mirror suspended in front of me all day long, and in it a Dorian Gray–like image, distorting me hideously after my failures. I was disgusted by what I saw; some days I didn't even recognize my likeness, so disfigured by incapacities I'd never known I had. It was at the end of those days that I'd fall into bed despairing, vowing to the dark that we'd never have another child because no one should be subjected to such an incompetent father. Privately—in the deepest privacy of my self, where I didn't have to admit what I was thinking—I dared to wonder whether I shouldn't see a therapist about my angry outbursts.

To distance myself from the turmoil, I craved time away, outside. When I wasn't looking after Carry or teaching, Griffy and I journeyed to places on the outer rim of the valley bowl. On the Palisades, directly opposite Cobble Hill, I found open-air freedom atop the cliffs. Tracing their curved lip as best I could, I looked for safe ledges where I could slide carefully out to the brink and peek timidly, breathlessly, down at the sheer walls below me, before settling back to inspect the valley from this new direction.

In early spring I learned my first lesson in microclimates on Sawyer Mountain, the next rise behind the Palisades, when Griffy and I seemed to pass through three different seasons on the same walk. Facing the river on a snowless eastside clifftop, I relaxed in the chilly autumnal shade and looked *down* at a soaring hawk for the first time; on the exposed southern bluff I watched tufts of emerald moss thaw in the hot spring sun and streak the rock outcrops with rivulets of meltwater; and heading down the snowy north-side path I walked back into late winter and then almost into a deer, who scared me to death when I accidentally startled it out from beneath a hemlock, its hooves exploding through iced-over pools as it burst uphill.

Some days later on Cottonstone Mountain, just across the river from Sawyer, in a spring squall that dropped fat snowflakes on already icy trails, I discovered that humility was not without its comic side. It happened on the way down, as I slipped and tumbled on the ice paths. No matter how slowly I went or how carefully I placed my feet, no matter how many trees I hung on to, I fell in every way imaginable—on my butt and hands and hips, elbows and knees—until my entire body hurt. I was embarrassed with myself at first, and then mad at the path. Finally, with the inadvertent help of the voice, who began a running commentary—"Oh, he's *down*! And he's hurt . . . He's writhin' in pain . . . It looks like that left elbow again . . . Can he get up? . . . Yes! He's up! Now how far can he go this time? . . . There's one foot out . . . and here's the sec- . . . Oh, no, down again! A bruisin' blow to his ass. That's gotta hurt!"— I was reduced to laughing at myself. Reluctant laughter to be sure, and defeated, but laughter nonetheless.

When I couldn't be alone, the landscape was even more helpful. One thing that soothed Carry was movement, and so on nice days I put him in the stroller or in a carrier on my back, and he, Griffy, and I revisited Burton's woods, exploring the changes caused by the main driveway and by the smaller, roughed-in roads that branched off toward the proposed house sites. Sometimes Rob would join us, and we'd all head out into the valley together to climb Houghton Hill or the Palisades or follow the forested seclusion of Potato Hill Road. In the quiet she and I would take a breath from the daily mania that had become our lives and actually talk about it, about how different life was now and how different we were, how many new things we were learning about ourselves, how much less time we seemed to have for each other. It was as if being out in the valley softened us and smoothed our edges. It allowed us to lower our guard: Carry's cries didn't seem as loud, Robin's worries less pressing, my self-disgust not quite as sharp. Life seemed more manageable, at least for those moments.

4

And over time, with no action in the development, our protest's edge softened too. Everyone had become so accustomed to the Uncle Asa sign that it was little more than a curiosity. The biggest question on our neighbors' minds was how long I was going to leave it up. Robin urged me to take it down.

But whenever I thought about walking out to the road and pulling up Uncle Asa's legs, I stopped. If the sign had lost its connection to the development, then unexpectedly it had found one to me. I felt like a birch tree I'd seen on one of my walks: risen high off the ground on its roots, it stood as if on fingertips, and held by those white rootfingers was a piece of slate—years ago an obstacle the root tendrils had to grow around, now a horizontal platform they clutched several inches off the ground. When I slipped my foot between two roots and stepped up onto the edge of the slate, it didn't budge. It held solid. In the same way this wooden structure, this sign born out of a goofy protest idea, had lodged itself unshakably within me.

I couldn't imagine the sign not being there. When Carry was born, a neighbor tacked a long blue balloon to it. When a snowplow ripped through one of its legs one night, I nailed it back together the next day. When I told people how to get to our house, it was always part of the directions.

And really, it wasn't just the sign. It was Uncle Asa himself. People would ask after him sometimes: "How's Uncle Asa doin'?" they'd say to me. One neighbor even took to calling *me* Uncle Asa. So the sign had gone beyond being just wood and nails and paint; it had evolved a life of its own.

I hadn't had a clear picture of Uncle Asa at the beginning. If he was part Asa Burton, he was also part fairy-tale character—a feisty spirit, a guardian saint of the woods. But over time, as more was expected of him, he filled in. Eventually I imagined him not as Asa Burton at all, but as that feisty spirit—an old farmer, a Vermonter whose family had come to Thetford not long after John Chamberlain had, who'd lived on this hill all his life, who farmed because his ancestors had and because he loved it, even though he'd never made much money from it. A widower grown beyond his productive years and a little beyond his wits, he'd sold off all but some chickens and spent most of his days puttering around the place. He was full of that libertarian Vermont spirit, not saying much unless asked, but then cursing the big hand of government and believing everyone ought to do whatever he wanted with his own land, as long as neighbors didn't have to see it.

And so he thought that if some rich flatlander doctor was going to buy the old Burton place across the road and chop it up so that other rich flatlanders could crowd onto the hill and build big houses, well, that was fine. But, by God, then he was going to finally make some money and do what no one else wanted to do, store nuclear waste and grow a fine breed of ticks. And he was going to advertise it too, right out there by the road.

But even this life history didn't quite do justice to Uncle Asa's character and the role he'd begun to play in my life. There was more to his presence, a real fullness I felt because of it, as if his persona had settled inside me. And it was sometime during the following months—with Carry turning one and settling into his life, with Burton Meadows looking like a piece of land speculation gone bust, with my growing sense of refuge in the valley—that I got a better idea of why Uncle Asa seemed so vital. I don't know what set it off exactly, but it may have been the arrival of some piece of mail from the Chicago area—the kind that would come periodically, say, from graduate school—addressed not to "Terry S. Osborne" but to someone named "Irving O. Seaman" . . . who happened to be me as well. Or had been. Or still was, sort of.

I was born Irving Osborne Seaman.

"Irving." The name came from my father through his father, making me the third in the line. That thirdness is what inspired the nickname "Terry." As did some compassion, I think: my parents must have known that "Irving" wasn't much of a name for a boy growing up in the sixties and seventies. And for the most part "Terry" was what I was called, until high school and college, when my new friends couldn't resist the temptation of calling me "Irv."

"Osborne." From my maternal grandmother, her maiden name. Reduced to "O" in the middle of my formal name. It was the least conspicuous of my three names and so the one I felt best about.

"Seaman." From my father's line. As time wore on, it became a primary source of my insecurity. Written, it seemed harmless—my high school French teacher, for instance, dubbed me *"le Marinier,"* "the sailor" —but orally it was a killer. From high school on my friends playfully referred to me as "Come" or "Jism" or "Sperm" or "Smegma," and so on. And at the end of college, on Class Day—the day before graduation, with our entire class of a thousand, and our parents, sitting on Cannon Green, behind Nassau Hall—I, Irving Osborne Seaman, was singled out for an award: The Worst Name in the Class.

There was general laughter and applause.

The voice had his own fun with that one. "You see?" he said after the event. "You're a fuckin' joke to everyone."

"I know," I said.

"A thousand people in the class, a thousand *names* in the class, and *yours* is the worst. The worst! The Worst Name In The Class."

"I *know*. Don't you think I know that?"

"Good," he said. "Just makin' sure."

By the time Robin and I moved to Chicago two years later, I had started collecting ways for avoiding such embarrassment. My older brother Peter, a screenwriter in Hollywood, told me that his name— Peter Seaman—was the source of endless hilarity for writers and actors and studio executives. One technique he'd used was to spell out our last name after saying it. I tried it; I became "Terry Seaman S-E-A-M-A-N" for a while, until it became too much trouble for the paltry results it was having. Pete also said that he would use his middle name for reservations at restaurants so he wouldn't have to hear, "Seaman, party of two, Seaman" called out over the intercom, while everyone in the lounge looked around to see who the loser was. I liked the idea of a new last name; it was what started me using "Osborne" as an alternative.

While I found ways to dance around new scenes of embarrassment, nothing was foolproof; there were always old scenes to be revisited, people from high school and college who still loved to joke about it. And there were other situations where my name couldn't be avoided. In the end, it was an Illinois Bell customer service operator who convinced me to change my name. I was on the phone, trying to correct my phone bill or something, and when I told the operator the name on the account— "Irving O. Seaman"—he began to laugh and couldn't stop. There was nothing to do but wait until he composed himself.

It was no different from the reactions I'd gotten for years. But this

48

one, I decided, was one too many. Up until then it had merely seemed silly to have a legal name I tried desperately not to use. But once that guy began laughing, I knew I would never use it again.

So I planned a new name. I would drop "Irving" and replace it with "Terry." And then I'd reverse the order of my middle and last names: I liked the way "Osborne" sounded at the end, and by putting "Seaman" in the middle, I could shrink it to a harmless initial "S" for formal situations and omit it the rest of the time. Presto: Terry S. Osborne. This was now who I'd be.

In creating the name, I didn't want to eliminate my birth name entirely and sever myself from the identity my parents had given me; I just wanted to minimize the embarrassing parts. And when I judged the new name from that perspective, I could feel good about it; I'd maintained two of my three birth names and added nothing that hadn't been attached to me before. But underneath I knew that damage was being done too, that the salvation of my ego was coming at someone else's expense: my father's. I was essentially erasing from my identity his entire name, which he had passed down to me.

I remember when I finally gathered the courage to tell him. We were alone, driving home one night from a hockey game in Chicago. I tried as gently and honestly as possible to tell him about the embarrassment I'd felt because of the name, but assure him that it wasn't any indication of how I felt about him personally. It was only about the words that made up his, and my, name.

He was hurt, I could tell, though he took the news stoically. He would support whatever I decided to do, he said, but he couldn't really understand why I needed to do it. He'd never had a problem with the name, and Pete seemed to be surviving. The difference, he thought, was that I had a "thin skin."

A thin skin. I'll never forget how jarred I was by those words, how little I thought he understood me. But as I look back on it now—my reaction to his comment, my reaction to *everyone's* comments, and finally my name change—I see the clearsightedness of what he said. And I wonder if my insecurity and self-deprecation, my panicked avoidance of embarrassment my whole life, my inability to ignore the voice, haven't all been intensified by the thinness of my skin.

I went ahead with the name change. The legal process took a while:

I had to hire a lawyer and post a legal notice in a local paper for six weeks and finally have a judge validate it. But it happened. And just a month before Robin and I moved east to a new place, I had a new name. If you'd stood back and looked at the sequence of events—seen me change my name and then get out of town—it might have seemed like I was running from something.

And if you'd looked again three years later, you'd have found me in that new place with that new name, still running, looking for safety in a valley bowl and yet another identity. Because what I'd discovered about "Terry S. Osborne" by then was that it was no more or less me than "Irving O. Seaman" was. It was just a name, and while it had erased a great deal of embarrassment from my life, allowing me to be less self-conscious in most situations, it hadn't dramatically changed who I was. When I woke up in the mornings and looked in the mirror, I still found a familiar reflection looking back, with familiar self-doubts and in-securities.

But Uncle Asa was different. He didn't care what other people thought of him. Or, rather, he didn't really think about it; it never occurred to him. He was not someone who could be embarrassed by his birth name or spring peepers, or who could be shouted down by a damning inner voice; not someone who could, while putting up his protest sign, happily imagine the stir it might cause and yet at the same time worry about the negative consequences of his actions, fear just as Robin had that he might actually *get in trouble*. And since he could be so untroubled, so different from the person I was inside, his existence gave me a freedom no name change ever could. Even better, he was somebody I could both be and not be whenever I wanted, somebody I could hide behind or proudly stand next to, somebody who'd lend me his best traits and absorb my worst ones. He had grown into an ideal alter ego, a wish fulfillment of the highest order.

5

Unfortunately, even fulfilled wishes run their course. Uncle Asa's time was bound to be up sooner or later.

Events turned on a crisp wind night in March. Driving up the road toward home, I came upon our mailbox lying outside of our driveway, a casualty of drunken sabotage by some local high school kids, who'd gotten to know the development in their own way during the last two years. They'd taken to using the recesses of it, well back in the woods where Long View Drive ended, for bonfires and drinking.

Asa Burton would have been troubled by the kids' behavior, but not surprised. When he first came to town, he found the young people here "ignorant as well as immoral, vulgar, & vicious . . . and excessively fond of vain, and vulgar amusements. . . . They made a practice of convening as often as every month for what they called a frolick, for singing & dancing, & rude behavior. . . . Between midnight & day break they would return home hallooing thru the streets, which often made me think of the howling of Wolves."

For all the time he spent civilizing the adults in town, he devoted even more to reforming the young people. He held weekly meetings, essentially Bible classes, in which he would entertain their questions about almost anything, from passages of scripture to rules of conduct and etiquette. The meetings "gained their attention, & furnished them with new thoughts, & served in some measure to correct their manners, & render them more attentive to preaching. Instead of wandering here & there on the sabbath it kept them more within doors to read the bible, & search for passages to be explained." By this format he kept them under his eye and began reclaiming their natures.

What *I* had to reclaim as I jumped out of my truck that March night was our mailbox from the middle of the road. I ran back to get it, and just as I was bending down to pick it up, I heard a crunching behind me, like boots on gravel. I thought someone was walking toward me. But when I looked up, I saw the strangest thing: my truck, driver-side door still open, was rolling away. I must have been so distracted by the sight of our felled mailbox that I'd forgotten to put the truck in park before hopping out, and now it was moving on its own toward the ditch beside our driveway. Mailbox in hand, I gave a couple of steps of panicked chase before stopping; there was no catching it. Off it went toward the ditch, door open, dome light on, warning bell dinging, headlights shining on the trees ahead. Somehow it managed to drive off the road not at an angle, which would have made it roll over, but straight and even, downward between three sugar maples, before coming to rest upright and unharmed at the bottom of the ditch.

But it had run down Uncle Asa along the way.

Now, while I'm no slave to superstitions, I do pay attention to them. I try not to walk under ladders; I won't open an umbrella inside the house. If I break a mirror, or if a black cat crosses my path, I feel a little unsettled and become watchful for a while.

I also pay attention to omens. I guess a part of me has believed all along that the world has its own meaning, separate from what we give it—a meaning constantly being communicated, and intelligible too, if you know how to listen for it or read it right. And when your own driverless vehicle gets it in its head to take off, and like a crack Mafia hit man breaks your protest sign off at the ankles, it's time for the sign to go, no matter how much a part of you it is.

The omen grew stronger the next morning, when I discovered that Uncle Asa and the mailbox hadn't been the only losses the night before. Across the street the Long View Drive sign had disappeared too; its stake was still upright, but bent and twisted like a corkscrew.

And one night some weeks later I heard through an upstairs window what sounded like gunshots coming from the development. Just another frolic for the Thetford youth, I thought. But the next morning I found the Burton Meadows sign split in two, the bottom half broken off. When I looked at it more closely, I could tell that the gunshots had really been the smacking sounds of someone swinging a baseball bat against the back of the sign.

I have to give it to adolescent vandalism: it accomplished in one night what I'd been dreaming of doing for two years. And it had done it even better, because instead of dropping the whole Burton Meadows sign legs and all, it had taken off only half of the sign board. Now just the top of it was remaining, and only one word—the name "Burton." Though I'd never thought of it before, that's exactly what the sign should have said all along.

Professionally, Asa Burton's life was a shining success. When he started his ministry in Thetford there were sixteen church members. Forty-five years later there were three hundred and twenty, "one of the largest churches in the state," he wrote. In addition he helped found the University of Vermont and was a trustee at Middlebury College and a leader of various regional committees and societies. And he was a sought-after speaker, giving guest sermons all over New England.

This success and popularity make even more striking the unhappiness evident in his writings. Part of that unhappiness was an understandable loneliness; he had outlived all three of his wives and two of his three daughters, and he mentioned repeatedly how grateful he was for the company of his last remaining daughter, Mercy, to whom he dedicated his autobiographical work.

But there was a different kind of loneliness in his words too, a bitter one stemming from betrayal. Burton felt that the townspeople of Thetford, the people he had devoted his life to, never showed him the kind of respect he deserved. Not only did his salary stay the same for over forty years, but he said he rarely received even that amount in full. Nor did he receive all of the firewood promised. He thought their behavior derived from their suspicion of him, their feeling that he was better off than he really was. It led them, he felt, to take advantage of him.

"My people have ever been deceived concerning my worldly circumstances," he wrote. "They have always viewed me as rich, tho poor in fact. . . . Instead of giving, & showing acts of kindness to me as a minister of X [Christ], they have called on me for money as tho there could be no end to my treasures."

This duplicity left him perplexed. "I know not how to account for this. For at the time they are drawing on my small income in every possible & lawful way, they profess great friendship, & regard for me, confessing I have been instrumental to their rising property & respectability in many ways."

It's hard to know exactly what happened, why the minister should have felt so abandoned by his congregation. But it seems possible that for all of the lessons Burton taught his flock, the one he could never get across, the one that may have seemed too strenuous for church members to follow, was the one about the interchangeability of outer conduct and inner worth. Perhaps when the townspeople saw Burton, stern and righteous, behaving like a prosperous man, when they saw the clean, ordered state of his barn and land—probably in better shape than most of the farms in town—they took the appearance of things, as most of us do, at face value: they saw his external worth as his external worth, and not as the glowing, moral face of internal nature. And so they judged him to be a much richer man than he was.

If that was the case—if Burton was misjudged, and even mistreated, because he so successfully lived out his principles—it was a brutal irony, a lesson on human nature as resonant as any parable. And yet even more brutal and ironic was the fact that Asa Burton himself never felt the comforting affirmation of his own inner worth. Despite his pastoral successes and exemplary conduct, he was no more charitable toward himself near the end of his life than he'd been toward the Thetford residents when he first arrived in town. "The longer I have lived the more convinced I have been of my own depravity, & ill desert," he wrote. "I feel more & more my imperfections, my nothingness, my weakness & dependence, & unworthiness. . . . I find, then, that a person's life may be filled with works visibly good & useful, yet the person performing them may have cause for self abasement, & deep repentance."

That final statement is to me the most jarring and tragic one Burton ever wrote. By recognizing that a person could improve himself, behave as well as possible, and still feel damned; or could improve the land, remake it in God's image—axe down the trees, roll them together and burn them, cultivate the earth, sow the seeds, harvest the grain and grass and corn—and still find something in himself wanting, unworthy, he was admitting the possibility that all of the principles he had been preaching and living by were unattainable, even for the most devout.

It was, I think, Burton's enormous despair—"melancholy," Freud would have said—that drew me closest to him. To have lived such a thoroughly pious life and yet have felt so worthless—how could that be? Well, I was starting to understand it. My life hadn't been nearly as pious as his, but fatherhood had forced me to notice that kind of darkness in myself.

So I used his words as a warning, a reminder, especially in the first months after Uncle Asa had gone. I tried to remember that any peace you made with God or the land or your child or *anything* was worthless without a separate, private peace inside. For no matter how beautiful the illusions you create in your life—whether a promise of righteousness or an ideal vision of yourself or a safe name or a private alter ego—they won't always endure. And like a moviegoer in a theater, the lights now brightening the darkened room after the show, you'll snap to and find that you're not really in that oversized world on the screen. Instead, you're like everyone else, a dreamer bound by gravity not only to the spinning earth but to the hard, uneven landscape inside your own thin skin.

Chapter 3

Mixed Landscape

I

Every morning begins broken, every waking with a tremor of consciousness. At best we surface lightly, buoyant as a bubble; at worst keelhauled, ripped from sleep by a towline, suddenly awake but airless, lungs small as raisins. And inside us the dark sea floor spreads, cracked land rubs against itself, our brain quakes. We're up.

September dawn and Carry is crying. Everything rolls on edge this morning—the molecular waves of his cry, the sun's blue light, the earth's autumnal yaw, our sleep. Carry's up. *What? What is it: ear infection? nightmare? hunger? new sleep pattern?* He's only two years old, so he can't really tell us. Even if he could, I'm not sure we'd get it. I mean, it's 5 A.M. Still, he's up, he doesn't care what time it is, and since it's my morning to be with him, I'd better start feeling the same way.

I look over at Robin. She's facing away from me, zonked. Doing her best impersonation of a rock, as I do on her mornings.

We had a hard night last night, she and I. One of our struggles about parenting, this time about how far to go in encouraging Carry to fall asleep on his own. We've been having to rock him at night while giving him a bottle, letting him sink off to sleep that way. But it's precarious. If we stop rocking or try to slip the bottle out too soon, even though he seems asleep, he'll snap awake, unhappy, and we'll have to start all over again. It gets frustrating sometimes.

Last night, we tried the "ten minute" method our doctor had suggested. We followed our normal routine, but once the bottle was done, we put Carry into bed, even though he was still partially awake. Then the trick was to wait for ten minutes to give him a chance to fall asleep, even if it meant enduring his crying.

Which it did. He probably would have cried the whole ten minutes if we'd gotten that far. But we didn't. By five minutes, Robin wanted to go into his room; I convinced her to wait. At six minutes, her face pained, she said she was going in, but for the next minute I guilted her into staying in the hall outside his door, telling her about his "best interests." At seven minutes, with both of us frayed by his frantic screams, we snapped.

"I can't stand out here any more," she shouted. "It's killing me!"

"Yeah, I can tell!" I yelled back. "But the question is whether it's killing *him*. And guess what? It's not," I added, though I wasn't so sure. "C'mon, Rob. I can't believe you're gonna go in there. That is *so* weak. You're gonna blow it."

"I don't care! What you're making me do now is cruel. Listen to him in there. How can you let him go on like that?"

"Oh, okay. So now I'm cruel. Thanks," I said. I could feel my anger building. "Hey, all I'm trying to do is stick to what we said we'd do. If we don't, he's not going to learn what we want him to. He's gonna win, goddammit!"

"Win?" she said, incredulous. "This is *not* some contest, you know. It's not a tug-of-war. What's the matter with you? He's only two!"

With that she opened the door to his room and went in, while I stormed down the stairs and out into the back yard.

Still a little resentful from that exchange, this morning I feel like making the bed creak so she'll stir and see just how early it is, how heroic I am, not cruel. But I don't. I quietly throw on some clothes as Carry cries. And as I stumble down the hall to his room, I can't believe how much of my life in the last two years has been carried out in utter fatigue.

It was the broken sleep more than anything that wrecked me at first, knocked all of my functions off-balance, contributed at least a little to my impatience. I felt as if I were driving in and out of unlit tunnels on a sunny day, going from dark to light and back again, not quite able to fix my eyesight, no matter how much I tinkered with things—sunglasses on then off, headlights off then on, visor down then up—no adjustment ever working well enough for my rods and cones to adapt, each change too quick to reverse the bleaching of their pigments. I felt darkblind first, then lightblind.

But slowly something happened. Just as the pigments, given enough time, do respond to the level of light, my sleep gradually found a home in its own disjointed rhythm. At some point I could spend an hour with Carry in the middle of the night—soothing, feeding, changing—and only vaguely remember it the next morning. Without knowing it, I'd been conditioned to exist in a middle ground; I'd learned to sleep while awake and stay partly wakeful in sleep. And it was in this ragged, fluid middle—a time between times, an alert space inside of drowsiness—that I began finding the best, uneasy rest.

When Carry started crying this morning, that zone was where I found myself, having fallen out of the early morning dreams I'd been dreaming—wild, last-minute scenes that screamed by like express trains, all rumbling noise then gone in seconds. Normally on early mornings like this one I take Carry downstairs and guzzle coffee to wake myself all the way up. But this morning I find myself carrying him out into the driveway, settling him into his carseat in the back seat of my truck, letting Griffy hop into the front, and getting in myself. Though I have no idea how they got there, two clear ideas are in my head: put Carry back to sleep so that he isn't exhausted the entire day, and drive some back roads of Thetford.

Strange. During my valley explorations these last four years now—all the while looking for some grand awareness, but in the meantime stuck learning the most fundamental things about the way nature works, things most other people seem to learn when they're kids—I've never set out in an adventurous, let's-get-lost-and-see-what-happens way. That's not the kind of person I am. I'm most comfortable if I know where I'm going, how long it'll take to get there, and when I'll be back. But this morning's different: I'm pushing myself into unknown territory. Or else something is pushing *me*. It must be that I'm still somewhere between waking and sleep: it's possible that, though I've stopped dreaming, I haven't quite awoken, or else I'm awake but still dreaming. I imagine myself as Gregor Samsa, the man in Franz Kafka's story "Metamorphosis" who wakes up one morning suddenly transformed into an insect, but who must have wondered just for a moment, when he first opened his eyes, whether the strange feeling in his limbs wasn't just the residue of undissolved sleep.

In the dawning September half-light, with Robin nestled peacefully in bed, unaware of our departure, Carry, Griffy, and I hit the road.

I start safely, as I always do, close to home, and drive to Potato Hill Road, which runs along the east and north sides of High Peak between East Thetford and Five Corners. It feels safe because I know this road already; Robin and I have walked much of it in sections, both while she was pregnant and then with Carry in a backpack. We've come to love it. But we've always worked from the East Thetford end and haven't ever made it to Five Corners.

This morning I decide to start from Five Corners, where at first the dirt road is flat and the area settled, with two long driveways on the right, two houses close on the left, before the terrain starts steeply up into the woods. After climbing over the piedmont ridge and descending back into the valley, the road becomes rutted, dropping downhill past the old Jenks place on the left, a beautiful camp with two ponds and three outbuildings. Then over the next rise there's the Sweet camp on the left, with wide views across the valley to Smart's Mountain in New Hampshire. I bounce slowly downhill in the empty quiet, seeing no one, down to the familiar pavement on Latham Road in East Thetford.

That was easy.

Next I drive west to Thetford Center and Whippoorwill Road, where I'd been once four years earlier with a friend named Butch, who gave me my first lessons in four-wheel driving.

"Roads like this, only one thing to remember," he said as I inched up to a puddled wheelrut. "Go!"

I never forgot that road, how remote it seemed, nothing but woods and a small cabin somewhere in the middle before emptying out suddenly onto Tucker Hill Road, right beside the house Butch was renting.

Driving it the other direction this morning, I can't believe it's the same road. For half a mile from Tucker Hill Road it isn't remote at all, but well inhabited. As I continue, though, there are fewer and fewer houses, then the cabin I remembered. Then for a stretch, only woods. After a time I come upon beautiful isolated fields on both sides of the road, their northwestern corners glittering in the sun. Then on the right there's a barn, then a house on the left, then a scruffy-looking pond on the right. Finally, I drive downhill and out to the pavement on Sawney Bean Road.

Hey, I know where I am!

Suddenly I'm feeling bold. I circle around, south to Thetford Center again, then head west to Rice's Mills—named after Samuel Rice, who operated both a sawmill and a gristmill there in the 1840s—northwest to Campbell Corner, and finally turn up Skunk Hollow Road, which I've never been on before. To the left I see a weathered saltbox house behind a screen of trees, and for whatever reason I have to get there. So, taking the next left, I follow the dirt road past a white mobile home, toward the house. While I'm imagining what to say to this family about my sneaking around their driveway at dawn—*Morning, folks. Driveway Inspector, Vermont Department of Driveways. Just making sure yours is in good order, which it looks to be. We'll put you on the Governor's commendation list. Thank you very much, folks. You have a good day, now*—I notice that it isn't a driveway at all, but a road that continues on past their house.

Thank goodness.

It leads quickly into nothing but shedding woods, becomes a leaf-covered track with no signs of recent traffic. After a little hill climb and a sharp turn left, the road descends slowly. For the first time this morning I hesitate, wondering if I'm going too far, beyond my limits, getting myself lost with my dog in the front seat and my now sleeping son in the back.

"Is this typical or what?" the voice says to me.

"What what?" I ask.

"You got no idea where you're goin', and you're gonna end up havin' to stop and ask somebody for directions, and when you stop Carry's gonna wake up and when he wakes up, he's gonna start screamin' and then *you* are gonna look like some third-rate kidnapper who couldn't mastermind his way out of a phone booth."

"Yeah, yeah, yeah. Whatever," I say. "Just wait." I've noticed stone walls on either side of the track, neatly lining it like a corridor. If stone walls are neat, I've learned, they're being tended; otherwise the succession of seasons would reduce them to jumbled mounds. So I'm guessing, hoping, that if somebody has been looking after the walls, that same somebody has also been using this road too, maybe as a farm road or something, and that it's going to lead us out somewhere.

Eventually it does, past a stunning spread on the right: a blue and gray stone house with a new sugarhouse and a barn-shaped guesthouse, a wide lawn sloping toward the road, apple trees studding the lawn. From there the road drops steeply; at the bottom I'm back out on Skunk Hollow Road.

"See?" I tell the voice. "You doubted me, didn't you? But the truth is, I *cannot* get lost. I am . . . Navigator . . . Man!"

"Yeah, yeah, yeah. Whatever," he says, mimicking me. Then he falls silent.

It feels great to shut him up like that; it doesn't happen often. What usually happens is that he just keeps talking, his verbal jabs filling time and space, keeping me unsettled. Or worse, when I'm most vulnerable, his words will jump me like playground bullies and wrestle me down and cinch my hands behind my back and climb on me to crow victory. He'll be yapping at me the whole time, and suddenly I'll feel angry and helpless, without the power to do anything that isn't granted by him.

It happened last night, as brutally as it ever has. Stung by seven minutes of Carry's crying and angered by Robin's accusation of cruelty, certain that his distress and her rescue of him were both my failing, I'd felt a shadowed heaviness coming over me. So I stomped down the stairs and out the sliding glass door into the night. In the back yard the dark closed around me. Not atmospheric dark; the voice's dark. Thick, hot, fetid. He was all I could see or hear.

"Nice work, loser!" he shouted. "Brilliant! Carry's still cryin' and Robin thinks you're torturin' him. Perfect. But you know what the worst thing is? You know what the absolutely most sickenin', disgustin', worms-out-of-your-mouth fuckin' disaster is? *You* couldn't stand the cryin' either. Actin' all strong. Tellin' her she was weak." He laughed. "Shit, *you* were weak!"

"No!" I tried to yell, as loud as I could, to show him that wasn't true, that I *was* strong. But what came out through my strained mouth was a thin, croaking shout that sounded as if I were trying to expel something rotten from my throat.

"Oh, *that* was forceful," he said. "Nothin' comes out the way you want it to, huh. You thought you could be an Olympic hockey player? Hah.

You didn't even make an All-Ivy team. A great writer? Sure. Your claim to fame is little do-it-yourself articles for a newspaper. Kick-ass college teacher? Right. That's why they only trust you with freshmen.

"And how about this perfect father bit? Strong and protective, sensitive and patient? A daddy and a mommy all wrapped up into one? Father of the Century? That's you, all right. Look at you, wanderin' around out here. Pathetic. You don't have a clue what's going on. You can't even get your kid to sleep!"

Marching across the yard in short zig-zags, it must have looked as if I was searching for something I'd lost somewhere on the ground. And, actually, I was. I wanted to find a release, a door out of that feeling, away from the voice and toward something more secure. Anything. But how was I going to find it when I couldn't see anything? There was only color around and inside of me—a blue-blackness thick as water. Eventually I made my way to the edge of the property where, beside a small lean-to that had once stored garden tools, I saw a glint down near my feet. Ah, that was it: the handle to the door I was looking for; maybe the latch to unlock it. One pull on it and I'd be outa here. But when I reached down, my hand knocked into something else, something hard I hadn't seen. What was it? Oh, a sheet of clear Plexiglas, rectangular, like a window pane. That wasn't what I wanted; the glint was what I wanted, and it was still there, flickering below the sheet. So I reached underneath and felt around until I touched the glinting thing and then pulled it. Out it slid, a long piece of chrome, U-shaped. It was a handle all right: an old lawn mower handle.

I held it up for a minute to look at it. Shit. No escape here. Then suddenly both my hands were on it, holding one end, and I was swinging it down like a sledgehammer. There was a sound of splitting Plexiglas. A good sound, the sound of something breaking. So I swung again for the sound, and then again, hammering the sheet into pieces, smashing down again on Carry's crying and Robin's accusation, again on all of my insufferable failures, grunting with every swing, sure it would start making me feel strong, again, but instead feeling worse and weaker, again, with every hit. And the color throughout me was darkening, and the darkness throughout me was thickening, and the thickness itself had teeth, those slivers of broken Plexiglas spinning around me again and again and again, like sparks.

Mixed Landscape

"Atta boy," the voice said. "Give it all you got."

Strengthened by my success so far this morning, I drive farther up Skunk Hollow Road across the Strafford town line, and take a left on Stevens Road, no idea where I'm going. The road winds uphill and down, with the same pattern of settlement—fewer and fewer houses—until Carry, Griffy, and I are alone in the woods.

Working slowly on the rutted road, I follow it sharply right and find myself somewhere else. Suddenly I'm riding along the side of a narrow valley in full fall color. It's as if I've driven ahead two weeks and into October; no one, no sound, just crimson hillsides and rising mist and sun. To my surprise the aloneness here has a different quality, not anxious at all. It's secretive, and comforting. Part of me wonders if I'm not setting myself up, getting a little too cocky by heading blithely into foreign ground, assuming that this valley road will have an outlet at the far end, like the others. The farther I drive on the road and the more remote it gets, the less likely that seems. But strangely I'm not so worried. In its remoteness it feels for that moment as safe as any place I've been.

Eventually the road cuts left, turning sharply out of the valley. At the corner stands a post with two hand-painted signs: one with an arrow pointing the direction I'm going—"Alger Brook Rd. Strafford," it says; the other points the way I've come—"Linton Hill Rd. To Miller Pond." At first I'm disappointed; the signs seem to signal the end of the morning's aimlessness, the return of the charted world. But then I feel myself relaxing: the words are a welcome sight, too. I imagine guardian sign painters looking out for homebody wanderers like me. The best part of having felt lost, I have to admit, is that moment of feeling found.

Once out of the valley, I bounce over a ridgeline and work slowly downhill. Halfway down, as I roll around a corner, I come upon the only creature I've seen all morning, a lone doe in her winter gray. She walks unafraid to the side and stops beneath a hemlock, looking back at me. I stop too and lean my head out the window.

"Get outa here! You'll get yourself killed!" I say loudly.

She flicks her tail, but doesn't move. Griffy, perked up by the truck's stillness, sees her and tries to step across my lap toward the open window. I drop my arm down so he can't; his nose, right next to mine, twitches to catch the doe's scent. And then we're all quiet. For a time the three of us just stare at each other.

"Well, it's a nice morning, isn't it," I finally say, softly.

She turns and bounds away.

I drive on. Eventually there are more houses and the dirt road becomes smoother and the smooth road empties into the concrete of the Morrill Highway, outside the village of Strafford.

This is familiar territory.

I turn around and look at Carry. He's out cold, breathing rhythmically, his head canted to the right against his car seat. Suddenly I have the urge to jump into the back seat and pull him out of his car seat and hug him tight and then tickle him into laughter the way I do sometimes. But there'll be chance enough to do that today: we'll be together for most of it.

"Okay, sweet guy," I say to him softly. "Let's go home and see Mommy."

Through South Strafford the back way, past the Elizabeth mine, where iron and copperas and copper were all mined periodically between 1793 and 1958. Over Gove Hill to Rice's Mills, Tucker Hill Road to Thetford Center. Route 113 home. Though Carry, Griffy, and I haven't traveled far, and though it's only been a couple of hours, I feel as if we've been worlds away and gone forever. But here we are still together, looking just as we did when we left. The morning is bright, the town is just waking up, and we're returning home at long last, safe and sound.

2

Even on our best days—when we feel awake, alert and perceptive, comfortable inside our skin—we're still artillery fodder in the trenches, the world pummeling us every moment with sensory stimuli. Though our sense receptors absorb what they can, there's too much to process, too much for us to feel everything, to perceive wholly and unchangeably at any one instant. In response, our sense organs and brain have made an extraordinary adjustment: they filter the stimuli. And so we perceive the world in layers.

Actually, we *are* in layers, sheathed in two layers of skin, as Diane

Ackerman points out in *A Natural History of the Senses*. On top is the epidermis—thin, scaly, and in a continuous state of slough. Below it lies the thicker dermis, containing our touch receptors. We actually feel with our second layer of skin, which is "why safe crackers are sometimes shown sandpapering their fingers," she writes, "making the top layer of skin thinner so that the touch receptors will be closer to the surface."

When it comes to adjustments, we're most familiar with the ones our eyes make: when we find ourselves suddenly in the dark, our photo-receptors respond, the rods' pigment slowly reconstituting: in minutes we begin to see what we couldn't earlier.

But our ears can adjust too. As Ackerman explains, the ear is constructed like a reflector that "takes sound and hurls some of it straight into the hole [toward the eardrum]; but a tiny fraction of the sound is reflected off the top, bottom, or side rims of the outer ear and directed into the hole a few seconds later." In effect, "we actually hear things twice," helping us to pinpoint the sound better.

And our nose acts like a snare, snagging odors in what scientists call a "neural net." In his book *A Second Way of Knowing*, Edmund Blair Bolles describes the way different neurons in our paleocortex fire during successive sniffs, giving different signals, even when we're smelling the identical odor. So the first sniff of chocolate may detect its sweetness, the next a scent of orange in it, the next some almond. Though we're unaware of it, the aroma has built itself in layers in our brain.

The roads I'd seen that autumn morning had struck me just the way a rich aroma might. I couldn't forget them—not only because it had been exciting to go places I'd never been and come out in other places I knew, but because I'd seen something, or thought I had. It seemed to me that the landscape had laid itself out in a distinct pattern, moving from fully developed ends through less developed midsections to a wild center and back out again in mirrored steps. The physical wilderness in town seemed bookended by civilized, settled areas. I wanted to find out whether that was true.

Following the example of the senses, trusting that a second look would lead to greater clarity, I went back to the roads. Over the next year or so I walked them all again, finding some of what I'd hoped: on each of the roads there was at least one place somewhere near the middle where I could stand, turn full circle, and feel surrounded by wilderness, see nothing but woods. And in every case the wild area was accompanied by a section of road that wasn't maintained or graded by the town, meaning it was rutted and uneven.

But aside from that everything varied. There were pockets of wild space, but no neat, common pattern; each road carried its interior wilderness differently. The untended area on Whippoorwill Road was a small middle piece; on Turnpike Road it was divided into two long sweeps that were separated in the middle by the junction with Houghton Hill Road. Potato Hill Road's wild space was more sporadic, broken by the seasonal camps on the east side. And Linton Hill Road's wilderness, though short, felt deepest of all, if only because it was farthest from home.

"Wilderness." I had to catch myself when I used that word. On my first drive along those roads, the middle sections had seemed like wilderness, at least the romantic ideal I had of it—enormous stretches of forest uninfluenced by humans. But when I saw those places again, saw that the bookending pattern was less perfect than I'd first thought, I also noticed that the woods themselves weren't as perfect either. Yes, these spots felt removed and untouched, but they were far from it.

The truth is, there's no real wilderness left in Thetford, Vermont. By 1850, ninety years after John Chamberlain arrived in town, three-quarters of the landscape had been cleared and turned to grazing and growing. The same was true for Vermont as a whole. Since there were so few settlers at first, and the process of clearing required the kind of physical labor that taxed Asa Burton's health growing up, the cutting of the town's forests probably began slowly. The most pressing need was to be self-sustaining: to build the necessary structures, grow enough food for oneself and one's farm animals, and provide fuel for cooking and heating. And while many trees were "wasted" in the clearing of fields— either girdled and allowed to die standing, or cut down and rolled into a pile with others and burned—some were used for specific purposes: basswood made good cabin flooring because its softness allowed logs to be easily hewn flat on one side; and Burton mentions that black ash bark was preferred for roof coverings, perhaps because of its thickness and flexibility. It also would have been important to leave a good deal of forest standing for fuel. In their book *The Story of Vermont* Christopher

Klyza and Stephen Trombulak note that a typical family farm in the early 1800s would have needed a forty-acre woodlot to provide a continuous supply of fuel. Given the fact that Thetford was originally divided into fifty-acre lots, we can make a guess at how a typical landowner in, say, 1820 would have apportioned his lot: about ten acres around the house and barn would have been cleared for agriculture while the remaining forty acres would have been left wooded for fuel and other needs.

But that formula didn't last. Population pressures and capitalism transformed the field-to-forest proportion dramatically in the following half century. The population pressure came from southern New England, where too many people were trying to farm land already depleted by agriculture. Many of them saw opportunity in the remote and relatively inexpensive Vermont landscape and headed north. Between 1800 and 1850 the state's population more than doubled, from about 150,000 people to over 300,000. Thetford's most dramatic changes came earlier in that period: between 1790 and 1800, its population grew from 862 to 1,478 people, and thirty years later it peaked at 2,113 people.

The capitalist pressures were more varied and longer standing, and probably more dramatic too, but they were inspired by a single cause: Vermont's inclusion in a regional economy. Again it was southern New England—its burgeoning industrial society, this time—that influenced Vermont's landscape most. Whether it was to harvest timber and float it downriver to busy sawmills, or whether it was to mine copper or produce wool or dairy products, Vermont began to play an important role as commercial supplier for the region, and its land began to be used commercially—that is, not solely to supply itself, but to meet the needs of people and industry to the south, primarily around Boston.

The result, by 1850, was a landscape whose field-to-forest proportion was completely reversed from the one fifty years earlier: now only one-quarter of the land was forested while three-quarters was open. The open land, devoted mostly to sheep during the wool boom in the 1820s, 1830s and 1840s, was being transferred more and more to cattle raising and dairy production, and the arrival of the railroad—in 1848 in Thetford—made the transportation of goods more efficient. This helped the area maintain its commercial viability through the time of the Civil War.

But by 1850 the railroad had made its way as far as Chicago too, and the country's primary attention began to turn west. The cheaper fertile land was there now, which meant new opportunity was too. And just as that opportunity had drawn people north from southern New England fifty years earlier, it drew people west from all over the east coast, including Vermont. Charles Latham Jr. notes in his *A Short History of Thetford, Vermont, 1761–1870* that there were 1,100 Vermonters living in California in 1850. Klyza and Trombulak write: "Emigration from the state increased, and the landscape of Vermont—which had supported native peoples for thousands of years—showed signs of being worn out less than one hundred years after the surge of European habitation had begun." Vermont was left to recover from its fifty years of economic and agricultural influence.

So even though the state's field-to-forest ratio has reversed once more in the last hundred and fifty years, meaning that three-quarters of the landscape is forested again, it would be a mistake to call these woods "wilderness." Too much has happened to the land. At best, what Thetford has now is *wildness*—swaths of second- or third-growth woods where pastures and fields used to be and where old-growth forest was before that.

While I didn't know the details of Thetford's landscape history then, I have to be honest: even in the wildest, most secluded sections of these roads, the places I felt most drawn to, whose purity I wanted so much to believe in, I couldn't ignore the reality. There was too much evidence of human disturbance. On the remotest spot of Whippoorwill Road, for instance, I noticed a run-down cabin not far into the woods to the west, and scattered debris—chain oil bottles, styrofoam plates, beer cans—on the ground to the east. You can stand in these lonesome places and hear only the wind. But if the wind is wrong, you can stand in these places and hear a chainsaw in the distance, or a single-engine plane from Post Mills airport buzzing overhead. On an east wind day on High Peak, the highest point in town, you can hear clearly the traffic on the interstate below, two varicose veins of pavement strung along the west side of the valley.

The more I let the illusion crumble, and the more I let myself see the woods for what they were, the more surprised I became. What suddenly seemed remarkable was that these regrown areas existed at all. It's true that the State of Vermont, by 1900, had begun to recognize the broader value of forested land, a value apart from its economic worth. That

worth, the state knew, would be best maintained by more farsighted forest management, and so it enacted and began enforcing forestry laws to
that end. But that effort was flying in the face of a less tangible but more
powerful American value: the moral distrust of wild space. The Calvinist belief in a fundamentally tainted soul—the kind of belief Asa Burton
held—inspired the early Euro-American disdain of untamed and un-
used land. Cleared land was more fruitful not only agriculturally, but
morally as well. "Civilizing the New World meant enlightening darkness, ordering chaos and changing evil into good," Roderick Nash writes
in *Wilderness and the American Mind.* Cleared land meant clean soul.

Later in his book, Nash excerpts Andrew Jackson's 1830 inaugural ad-
dress: "What good man," Jackson asked rhetorically, "would prefer a
country covered with forests and ranged by a few thousand savages to
our extensive Republic, studded with cities, towns, and prosperous farms,
embellished with all the improvements which art can devise or industry
execute!"

Americans seemed to agree with Jackson, because we cleared the land
from coast to coast and in the process, of course, cleansed our souls from
top to bottom. Rid ourselves of all evil. Except, once the country's land
had been cleared, an interesting thing happened, similar to what happened in Thetford after 1850: settlers suddenly noticed what it was they'd
been fearing. As Alexis de Tocqueville remarked, "[Americans] may be
said not to perceive the mighty forests that surround them till they fall
beneath the hatchet." Not only that, but Americans didn't feel quite as
saintly as they imagined they would either. In fact, quite the opposite
happened: they felt an increased urgency, as if something were pressing
in on them, or else getting away. Part of that urgency was economic, to
be sure: the agricultural boom times were ending. But part was moral as
well: to everyone's surprise, wild space and all its dark force became
more threatening in its absence than it had been in its presence, and
people devoted just as much time, if not more, to keeping the woods at
bay as they had to clearing it in the first place.

In Thetford, Fred Howard, who still owns twenty-five acres of land
that his father bought on Turnpike Road in 1900, remembered, "My
father used to say, 'If somebody doesn't get busy with an axe, this country will be a howling wilderness before long.'" Those words carry all
the moral anxiety of a Jonathan Edwards sermon, or even one by Asa

Burton, voicing a fear not only of a republic in decline, but of mortal sin and impending damnation too. As late as 1947 Fred took an axe with him up Turnpike Road to try to keep the pasture clear.

Feeling the moral weight of something only in its absence—an absence you've caused—is American nearsightedness at its best: Tocqueville's disdain was well deserved. So then why, a century and a half later, are the woods back? Is it simply a matter of economics again, the result of a shift from an agricultural landscape to a more service-oriented, residential one? Or does it have to do with something else as well, maybe with a shift taking place in our own psychological landscape? Perhaps we recognize, in ways we're not even conscious of, that it's easier to live amid a threat we can see than one we can't. Or maybe we're beginning to sense that the woods are no more or less a threat to us than our own internal demons.

I had hoped those repeated walks on the roads would provide the answer, but they didn't. At least not right away. And the voice picked up on that.

"So, what's the matter?" he said.

"Nothing," I said. "I'm doing fine."

"Like hell you are. You're lost. Remember that morning you were drivin' around these woods all warm and fuzzy, thinkin' you could find your way anywhere? You know what happened, right? You got lucky a couple of times is all."

"That wasn't luck," I said. "I knew we were fine."

"Oh, yeah? Well, here y'are, know-it-all. You come back here and it might as well be the moon. The roads aren't like you thought, the woods aren't like you thought. You think you're in your private little Shangri-La out here, alone in the forgotten forest, but there's trash everywhere and houses around the corner and cars goin' by. Boy, you had it figured. Real wilderness."

"Hey, I'm still learning about the woods, okay? It's going to take some time."

"If I were you, I wouldn't even bother," he said. "You won't ever understand 'em. Just stick to being Little Dreamer Boy, pedalin' your toy truck around the hillside. That way you won't have to pretend to know things you don't, and you won't end up lookin' so stupid."

Despite the voice's chatter, I did, slowly, begin to understand the changed feeling of the woods this time around; as it turned out, it had less to do with the appearance of an interior wildness than it did with me and the quality of my perceptions. Something, on my second and third visits to these places, seemed absent from the experience. For the longest time I couldn't figure out what it was, but finally it dawned on me: it was the Little Dreamer Boy. *He* was missing, and so was the magical feeling of discovery that had been with him that September morning in his truck.

Part of this, I think, can be explained by the senses and the conditioning of our perception: our brain reacts most strongly to new sensory stimuli. Ackerman writes:

> There is that unique moment when one confronts something new and astonishment begins. . . . but the second time one sees it, the mind, says, Oh, that again, another wing walker, another moon landing. And soon, when it's become commonplace, the brain begins slurring the details, recognizing it too quickly, by just a few of its features; it doesn't have to bother scrutinizing it. Then it is lost to astonishment, no longer an extraordinary but a generalized piece of the landscape.

But perceptual dullness, caused by wandering the roads over and over, wasn't the only factor. It occurred to me finally that intention was the larger problem: I was *looking* instead of seeing. Going back deliberately the way I had, trying to find exactly what I'd seen months or even a year before, was like trying to reenact consciously the drama of a recent dream: just the act of trying guaranteed failure, betrayed me as too much of a literalist, and maybe even too naïve, for my own good. I'd missed the most powerful quality of the original experience, those dawn perceptions: they were uncopyable. I certainly couldn't recapture them on purpose.

Disappointing as that was, perhaps there was a reason for it. Edmund Bolles notes that while the brain's motor system functions normally during dreaming, certain nerves block the brain's messages to our muscles when we're asleep. If they didn't, what we'd have is what French researchers discovered when they cut those nerves in cats: cats bounding out their dreams while asleep. But even more important, without those nerve blocks the same cats moved much more impulsively, much less cautiously, while awake. In other words, they leapt before they looked, which isn't very conducive to species survival. The point must be that our

systems are designed to keep us from interchanging levels of consciousness deliberately, of crossing them too easily, of perceiving the world
wholly every second. It's difficult to do, in part, because it's suicidal.

When it came to my explorations and research on the roads, I began
to recognize different varieties of perception and notice the way my de-
liberate, conscious attention often dulled it. The sharpest, most enduring perceptions were the ones unlooked for, the ones I simply happened
upon.

One day I visited the Thetford Town Hall to see if Arthur Bacon, our
town lister then, could give me any background on the roads. He said
casually that he might have something, then went behind a door, pulled
out a suitcase and laid it on the table. Opening it, he pulled out some
papers. It turned out that he'd researched the history of all the roads in
town and kept the records in that briefcase. He told me I was a little
confused about Whippoorwill Road; it didn't go all the way through
from Tucker Hill Road to Sawney Bean, as I'd thought, but merged into
and became Poor Farm Road somewhere in the middle. Poor Farm
Road had gotten its name from the farm the town ran between 1840 and
1890 to give the town's poor a place to live and work. And Wells Fargo
Road, which I'd thought was that driveway I was sneaking into off Skunk
Hollow Road, got its name because Henry Wells, who co-founded the
American Express Company and Wells Fargo and Company, was supposed to have grown up near there. (Arthur thought he'd found the
house's cellar hole in the woods.) The earliest mention he could find in
the town records of Turnpike Road was 1803, Potato Hill Road 1811,
Whippoorwill Road 1812. He knew that Poor Farm Road, under a different name, was laid out in 1808. And on and on and on. As he flipped
through his pages of notes, I stopped listening for a minute and just
watched Arthur himself, awed by how much he knew.

Another day I saw the roads from the air, at the Soil Conservation
Service office in Randolph, Vermont. I hadn't really meant to. I'd gone
there to learn how to use a stereoscope on aerial maps. After helping me
with the stereoscope, Bill, a geologist there, laid out an enlarged infrared
aerial photo of East Thetford taken five years earlier, in 1987. In infrared,
the landscape looks different: green fields are red, bare soil is turquoise,
the Connecticut River smooth black. Disorienting until you get used
to it. Looking down at the roads strung across the terrain, I could see

clearly the areas of interior wildness on Whippoorwill and Poor Farm and Potato Hill roads; so open and developed at their edges, the roads disappeared from view further in. But Turnpike Road, probably the least developed road in town, was remarkably visible from end to end, uphill along Gun Brook, then snaking back downhill to Five Corners, woods bordering its whole length. I could never figure that out.

And during my walks on the roads themselves, it was each surprise encounter, not the wildness itself, that imprinted deepest: on Poor Farm Road I saw a scarlet tanager for the first time, redder than I imagined any bird could be; on Wells Fargo I happened upon an antlerless buck that sprang uphill when it saw me, but somehow left no hoofprints in the road's spongy mud where I could've sworn it had stood; there was the stem of white baneberries on lower Turnpike Road that made me wonder if I hadn't come across an alien atomic structure, the diaphanous rose light inside the stand of red pines on upper Turnpike, the gold corona around a hawk as it glided silently across the sun above the old Jenks place on Potato Hill.

We look at one thing but glimpse a much more vivid flicker out of the corner of our eye. We give up listening and suddenly a sound swirls around the rims of our ears, drops inside, and stuns us. Leaning intently over a flower, we sniff, making new neurons fire, and we catch a scent the wind has carried across the ridge. Our sensory experience seems to defy our intention. It's like trying to grab a floating dandelion seed: the wave of air we make while grabbing just pushes the seed away. We lunge and lunge and lunge, sending the seed farther out of our grasp. The problem is, each time we open our hands, not finding the seed, we look ahead for it again, convinced we've caught nothing.

3

The summer my older brother Peter was seventeen, he got out of bed one night, pulled on a pair of pants and walked shirtless and barefoot out to the garage; once in the car, he backed out of the driveway and drove off down Elm Tree Road. At the corner of Elm Tree and Westminster, he rolled through the stop sign as he always did and headed west on Westminster to its four-way intersection with Sheridan Road, where he stopped. And try as he might—looking left, right, straight ahead—he could not decide, sitting there at the stop sign, which way to go.

So he woke up.

I remember, at age nine, being stupefied by the story. How did he do it? Did he have his eyes closed, or open? Why couldn't he decide which way to go? How did it feel for him to wake up and find himself sitting there at the intersection?

None of Pete's answers could live up to the astonishment of my questions—he'd been asleep, after all, and hardly able to make sense of it himself—so I just stored the episode away, added it to the mythology of this older brother I already idolized.

Our mother reacted differently. The incident terrified her: she pictured her son asleep at the wheel—unconscious and out of control, as if having nodded off from fatigue. For a long while she didn't sleep well herself; never more than half-drowsing at night, she stayed alert at some level for the sound of a car starting. It was finally our family doctor's reassurance that helped her most: Pete, sleep-driving, was fine at the wheel, he told her. Safer than any of us is awake. He said: trust the dreamer's instincts.

It was this advice that echoed back to me as I wondered about Thetford's landscape. I believed what I'd seen. Though I couldn't recapture the dreamlike clarity of that first morning on the back roads, I knew that the wild spaces were there. But I couldn't figure out why. And I didn't understand why they seemed to be bookended across the terrain.

I asked Charles Hughes, who owned three hundred acres of land on the east side of Poor Farm Road. It was his barn and house and secluded fields I'd passed that first morning just before coming out on Sawney Bean Road. He'd bought the land and farm after the hurricane of 1938; of the fifty acres that were open when he bought it, only ten still remained as pasture, though he had tried to keep clear as much as possible.

"I'd have it forest if I had to choose," he said. "But there's so much cleared land that's gone back to forest that I've felt an obligation to keep some of it open."

Fred Howard's family grazed cattle on their Turnpike Road land until the 1950s; after that, it wasn't profitable. And without cows grazing it down, it was more trouble than it was worth to keep it clear.

Fred said he'd prefer to have the land open, if it served any purpose.

"Sure I'd like to see it in fields," he said. "But I'd rather see it in trees than in houses. A lot of good agricultural land has been transferred into housing."

One man, preferring wooded land, kept as much open as he could in deference to history. Another preferred cleared land but left it wooded because there were worse alternatives. If the two were representative, then Thetford's regrown wildness may have resulted from cross-purposes, or best intentions gone awry. Or at the very least, by accident. It had become what Middlebury College professor John Elder calls an "inadvertent wilderness."

Around here any open land left to itself will regenerate, turn to woods, go wild. That field-to-forest succession might happen in any number of ways, but in general it would begin with pioneer species, species that grow well in the open, such as white pine or quaking aspen or pin cherry, followed by successional species such as yellow birch or sugar maple, white spruce or hemlock, trees that can thrive in shady conditions, waiting for their turn to replace the pioneers. But so much depends on circumstance: how the open land was used and for how long; what trees are nearby; what the surrounding climate is. For instance, Tom Wessels in *Reading the Forested Landscape* points out that a long-overgrazed pasture might be pioneered by common juniper, which, when grown to shrub size, would play the role of protective nurse shrub for any number of tree species whose seeds found their way beneath its branches, leading to a very mixed young forest. But in a pasture less heavily grazed, where juniper hadn't had the chance to take hold, the new forest's composition would more likely be dominated by a single species—one that happened to have its mast year, its heaviest seed production, at the right time.

In a designated wilderness we purposely leave these processes to themselves. But wild areas that seem to have grown up inadvertently, like those around Thetford, are so much more provocative, and so much more telling about us, who've been inadvertently responsible. Like a searing dream that bursts on us out of nowhere, these wild spaces threaten to blow our cover to ourselves, reveal more about us than what we do on purpose, show us things about ourselves we don't necessarily want to see—such as our involuntary attraction to wild space, which we've long disguised as moral disgust.

Edward Abbey understood that. He wrote, "We need wilderness

whether or not we ever set foot in it." That's a hard concept for Americans to accept, because we've spent so much time clearing the land deliberately, civilizing it, making it productive for agriculture. Making ourselves feel civilized and productive too.

But at some level we must know we're not like that; our deepest interior is probably as unpolished and wild as the land's. What if some day our attraction to wilderness, our need for it, grows so strong that it finds its way out through our civilized shell? What would prevent our inner wildness from reappearing as inadvertently as these regrown woods have?

Nothing, as far as I could tell then. If it could happen in this valley, it seemed inevitable inside us. As Carl Jung wrote, "In [our] civilizing process we have increasingly divided our consciousness from the deeper instinctive strata of the human psyche. Fortunately, we have not lost these basic instinctive strata; they remain part of the unconscious, even though they may express themselves only in the form of dream images."

What, then, was I supposed to do? Trust the bookended wild areas the same way I trusted my dreaming instincts? Both exposed an inner wildness that was always there. And if I did trust them, what did that mean about me, about the voice and what he was and who I was? I'd only ever thought of the voice as a natural part of me; he'd been with me for as long as I could remember. But now I had to wonder whether he was something in and of himself and whether he was supposed to be there or not. Was he simply a wild space inside me, necessary and essential? Or was he something else, something alien and excessive—a wildness that had outgrown and distorted its allotted space? Or maybe even a wildness invading a foreign place, choking the native system there, like a desert having formed where woodlands should have been?

Sometimes the voice felt like that—so separate from who I thought I was, so strange and destructive, so much not like me—and yet in arguments with Robin I'd hear his voice coming out of my mouth, blaming her for everything he usually blamed me for, shouting her down until she gave up, exhausted and discouraged, the way I did under his relentless pressure.

After one of those arguments, days later, when we'd both cooled down, she said, "I've been thinking about us. How much tension there is between us, and how unhappy you seem. Whenever you have free time, you go off by yourself on your walks. You don't even tell me where you go.

I get worried that it's me you're angry at. But then I see how impatient you can get with Carry, and I start to think it must be something else."

"But you can get that way, too," I said.

"I know. It's been hard for both of us. We both want to be such good parents. But it's made me notice how different you are from the fun, outgoing person you were in college."

"Wait, wait, wait. That's not fair. People change. I don't have to be the way I was in college, do I?"

"Of course you don't," she said. "I'm not saying you do. I've changed too. But I was watching you play with Carry one day, and you just looked so sad and tired. So I started thinking about it, and I could sort of see a pattern, you know? I don't know exactly when it started, but it seems like your down moods have been coming and going for a long time, even before Carry was born. So, I was thinking, maybe it wasn't just a change in moods or a change in you."

"Yeah? So?"

"Well. . . maybe you have a low-grade depression or something. You just don't seem happy with your life."

"Is she fuckin' kidding?" the voice started screaming at me. "Depression? That only happens to nutcases! What does she think you are, a psycho? Shit, you're not that bad. You just get stressed out every once in a while, that's all. Low-grade depression, my ass."

I can see now how unthinkable it would have been for me to embrace Robin's suggestion, though somewhere inside I was scared she was right. The voice controlled what I saw and how I saw it, all of which favored him. And of all his characteristics, that one—his ability to preserve himself by explaining or denying or arguing himself away—is the one I still marvel at. Why was I so quickly convinced by him and not at all by Robin, whose doctorate in clinical psychology and experience as a therapist made her a much more reliable source? In *Darkness Visible*, William Styron's memoir of his depression, he offers an explanation that rings true to me. Noting how willfully ignorant he had kept himself about depression while learning a great deal about other medical topics, he writes: "Most likely, as an incipient depressive, I had always subconsciously rejected or ignored the proper knowledge; it cut too close to the psychic bone."

And so, afraid of being discovered, I steered Robin away.

"Really? God, I don't know, Rob," I said calmly, trying to sound open to the suggestion, hoping my openness would make her drop it. "Life seems okay to me right now. It's just harder than I expected to be a father, I guess. I don't know."

And the truth was, I *didn't* know. Maybe it was more than that. Maybe I *was* depressed. Or maybe I was crazy and really there were two of me and I couldn't control which I was going to be from day to day. Or maybe the voice was really who I was and I was the disguise, and up until now I'd been able to cover him over, but from here on in he was just gradually going to replace me with the blue-blackness I couldn't see through. How would I know?

As I thought about the dawn drive and the retracing walks I'd made this last year, I could see more clearly what was happening: I'd begun trusting the land for those answers. And that vague longing I'd felt for the view years ago had now distilled into a hope: that on some new morning in the ebb of a dream I'd see in the terrain my full self, a self I could understand. And then I'd know who I really was. Part of me was afraid of what I'd find, that it wouldn't be pretty, not pristinely wild like the white pine stands that Europeans first encountered here, nor smooth like Thetford's fields and pastures in the 1840s. But if it wasn't pretty or distinct or a single entity I could make sense of, at least it would be, like the rest of the world, a mixed landscape—clear and groomed in some places, remote and unkempt in others.

"Were the earth smooth, our brains would be smooth as well," Annie Dillard writes. "We would blink, walk two steps to get the whole picture, and lapse into dreamless sleep."

There were times when such numbness sounded comforting, worth seeking out and settling into. But then I'd step out onto the back deck and see the unevenness out there—the wooded ridgelines rippling north, the valley slopes riding down from them, the cleared bottom land beneath it all—and know that numbness was not in my future. And though that recognition scared me to death, I wasn't about to turn back. I didn't want to turn back. A state of smooth numbness would have made things easier: with me perceiving everything at once, the journey itself would have been unnecessary. But then I'd have lost the chance to learn anything at all.

Part Two

Water

Chapter 4

The Water Drill

In a letter he sent me a year before we moved to Cobble Hill, Richard Ford, a brilliant writer, and a teacher of mine in college, wrote, "One's education is inevitably slow dawning and often not simultaneous with time spent in school." I remember being drawn to that sentence immediately. I even read it aloud to Robin. Not only did it sound beautiful, but it seemed to carry something important: a blessing, a warning, an emollient, a key to a locked code. I didn't quite know what.

I could have used its wisdom right away, on my walks around Cobble Hill and those early journeys into the valley when, stretched tight as piano wire, I found myself strangling knowledge out of every experience to try to quiet the voice's mocking. But it wasn't until I crossed over the Vermont piedmont and explored the Ompompanoosuc River, a tributary of the Connecticut, that I understood the idea better. Along the Ompompanoosuc's east branch, between Great Falls in Thetford Center and its headwaters in the town of Vershire, I recognized more and more the potential for a different contour of knowledge, a different rhythm of mind. It was a rhythm no schooling could amplify, and no urgency alter, embedded as it was in the work of water.

Though I didn't know it at the time, the recognition process had begun even before I took to the rivers; it had started earlier that year, on the day my good friend Ned Perrin and I tried to thaw a frozen culvert beside his sugarhouse. Actually, if you wanted to trace it further, you could go back to the day Robin and I moved onto Cobble Hill and found stacks of dented sap buckets and piles of rusted taps in the old sugarhouse in the northeast corner of our yard.

Built just after World War II, that old sugarhouse originally belonged to the Evanses, our neighbors to the north, who'd farmed the land on this side of Cobble Hill since 1938, before selling much of their property in the early 1970s to the people who built our house. To make room around the proposed house site, the new owners moved the sugarhouse about seventy-five yards to the east, setting it in the shade of a big white pine. But they still used it, still boiled sap in it every spring, as the Evanses had before them. Fifteen years later, the property was sold again and divided up, and by the time we came along our house stood on a three-acre lot that had only five mature maples on it. But the sugarhouse was still there, considerably aged now, its lower end propped precariously on cinder blocks, part of its roof rotting through, its windows covered with Plexiglas, and inside only a pile of worn buckets and taps.

Ned, who had his own sugaring operation at his place, was the one who got me started. He scouted out the five maples around our house and found eight others at the top of our driveway that weren't actually on our land but that our neighbors didn't mind me tapping. Then he lent me some taps and taught me how and where to drill the holes into the trees, and with some usable sap buckets I salvaged from the sugarhouse and others I picked up at a yard sale, we were in business. As much as our schedules allowed, we worked together, tapping trees and hanging buckets, collecting the sap and transporting it to a holding tank outside his sugarhouse and feeding it into the evaporator inside. Then, in a fog of sweet maple steam we boiled the sap down to syrup.

If sugaring fostered my friendship with Ned, it also built a bond between Carry and me. Even before he could walk, we'd sugared together. I had put him in his backpack and taken him outside with me while I drilled the holes in the trees and hung the buckets, or later in the season collected sap from them. Then, last spring, just before he turned three, he joined me on his own feet. He loved to clean out the holes after I'd drilled them, poking a twig in and swirling it around the way Ned had shown us, to get the last curls of wood out before I hammered in the tap. And if the sap was running, we'd stand still after hanging a bucket and listen to the hollow *pings* of the first drops hitting the bottom.

"Huhhh?" I'd say, making an inhale of surprise, our way of getting

each other's attention, asking whether the other had heard some familiar noise—a tractor's grumble, a woodpecker's rapping.

"Huhhh?" he'd inhale after another *ping*, to let me know he had.

I wasn't sure exactly when this had happened, this reciprocal interaction between the two of us, and the coded communications we'd developed, but I knew it had been critical to my settling in to fatherhood.
Maybe it had begun as early as his infancy, when I'd answered his discomfort with a course of Boober Doobers and he'd answered back by
gradually quieting. If that was so, I wished I'd been secure enough to recognize it. Because at the time, that had been the problem: I didn't know.
I didn't know whether Carry knew who I was, or how hard I was trying,
or how bad I felt that I couldn't make him feel better. Actually, for the
first months it was clear he *didn't* know who I was. Or care. And why
should he have? He was an infant. Still, that initial anonymity was
something I hadn't expected, and because it seemed to confirm all of the
voice's dark predictions for my life, I'd responded defensively by treating
Carry more like an adversary than a companion.

But now there were moments like the one amid the sap buckets,
when he showed how well he did know me—knew who I was, knew
what I was trying to tell him—by answering my intaken breath with
one of his own.

Separated from the Ompompanoosuc by a bankside clearing and a driveway, Ned's sugarhouse stood fifty yards uphill from the head of Great
Falls. Close by, no more than fifteen feet away from the door, was a
brook that began further uphill in some swaled upper fields, cutting diagonally through their center like a crease in a cupped palm. From there
the brook ran downhill through a swamp, past the sugarhouse, through
a culvert underneath the driveway, beside the small clearing and finally
into the Ompompanoosuc. Even in the driest months it trickled noisily,
and harmlessly, in its trough beside the sugarhouse. But spring was its
boom time, as it drained the thawing fields.

Of all the seasons around here, spring wore the slow-dawning idea
best, not just because of its characteristic brightening—the sun rising
toward its summer zenith, the days lengthening in response—but also
because of its unhurried secrecy. Like a clock without a second hand,

spring's progress was elusive but inevitable, coming so slowly you didn't notice it until sometime in May, when you turned around and found the rhododendrons in bloom, the grass long and emerald, the maples in full leaf. And this chlorophyllian miracle seemed to have taken place in an instant, overnight.

But it hadn't. What you'd missed was the undetected work of the last two months, the way the grass had emerged from its winter cover pale and splotchy, like sickly skin. You'd missed the way the soft maples turned spidery red, the sugar maples red then amber, the birches a neon lime green. You'd missed the myriad changes in the weeping willow beside the Evanses' house, from cream to mustard to wheat to bright yellow, a different color almost every day. You'd missed the front-yard oak holding and holding and holding its auburn bloom until every other growth had gone green. It had been a long, subtle process, this budding, though it took you by surprise.

That March, spring had a surprise for Ned and me when it brought a handful of warm days after a long cold spell and set the fields to thawing. We'd been preparing for sugaring in a relaxed way—tapping a pipeline in one stand of maples, hanging some buckets in another. One day we loaded the evaporator's flue pan into Ned's truck, planning to haul it into the sugarhouse. But when we got there the building was almost inaccessible. An island in standing water.

After investigating, we saw what we were up against: it was the culvert under the driveway. Thirty feet long and about four feet in diameter, it was frozen solid, and was acting as a dam against the brook, which was trying to drain the upper fields' runoff. Our first reaction was primitive: fight or flight. Since flight would have meant a flooded sugarhouse, we chose to arm ourselves with tools of destruction instead—an axe, a sledgehammer, an iron bar. Then we returned to the downhill mouth of the culvert, where ice bulbed out like a swollen tongue, and we attacked it.

We swung.

We bashed.

We chipped.

Into every noticeable dent or crevice we threw a handful of halite to speed the melting, then swung and bashed some more.

Here's what half an hour of fury earned us: sore shoulders, blistered hands, a dulled axe, and twelve inches of headway. It may not have been productive, but it felt like it was, at least to me, because it was a kind of brainless, bludgeoning response I could do as well as anyone.

Still, we needed something better, and while we rested, Ned came up with an idea. Instead of trying to blast open the original channel, why didn't we do what the water would do, if it could? Why didn't we create a new channel of less resistance?

Back at his barn we found fifty feet of black one-and-a-half-inch plastic pipe, which we laid downhill from the edge of the floodpool. It took some time and a good deal of sucking on the pipe to start the water flowing through it, but eventually it flowed. A solid fast stream. We had diverted the brook with a siphon.

I was thrilled with our work; Ned looked at it dubiously.

"Risible," he said.

I looked again, and then again, and finally saw his point. If you'd walked uphill to the huge floodpool, then come back downhill to that little fingerflow of water, you'd have been tempted to laugh.

But we didn't. It was all we had.

Then Ned came up with one more idea. He wanted help maneuvering the downhill mouth of the pipe to the downhill mouth of the culvert, in order to direct the draining water toward the bulbed ice.

"How come?" I asked.

He said, "Running water, no matter how cold, thaws ice."

We dragged the pipe over to the mouth of the culvert, careful not to yank the top end out of the pool, and fixed it in place so that it splashed its stream against the ice face.

I was stunned; it was ingenious. We'd made double use of that single pipe and let the flood resolve itself: as the water drained from the pool above the culvert, it drilled away at the ice dam below it. We'd turned that pitiful stream into a one-and-a-half-inch water drill. I felt as if we'd invented the wheel. In a better world, where ingenuity guarantees success, that would've been the last of the ice dam and the end of the story. But victories of ingenuity are often fragile ones, wildflowers half-blooming from the dry ground of a larger defeat. Ned checked the culvert two or three times that night. The pipe was still draining the water and drilling the ice, and the runoff had slowed with the cooler nighttime temperatures, but there were no dramatic results. The floodpool still looked like a small lake, and the water drill was simply smoothing out the rough edges we'd chipped into the ice tongue. It was going to take our drill a long time to thaw that blockage. By morning, when it was clear that another warm day would lead to shin-deep water in the sugarhouse, he called the Thetford road crew for help.

One of the crew arrived with the town's culvert-steaming rig. It was a metal drum in which water, heated by propane, was converted to steam. The steam passed out of the drum, through a long wand of half-inch galvanized pipe and out a pointed, perforated end. Using the steaming pipe as a lance, he prodded the ice, boring two fist-sized melt-holes through which the water could flow and finish the thawing itself. It took an hour or so but it worked. The next day we were boiling sap in a dry sugarhouse.

Afterward, whenever I was around Ned's place, I couldn't help thinking about the water drill, about the triumph of inventing something from limited resources, though the invention ultimately failed. It seemed something worth remembering.

Then one day I asked Ned why he'd been so nonchalant about his invention. He told me that he hadn't invented anything. As far as he knew, water drilling had been a mining technique for a long time. When I checked it out, I found he was right. In 1853 near Nevada City, California, three miners—Edward Matteson, Eli Miller, and Anthony Chabot—collaborated on the first mining water drill. Using an uphill reservoir, a canvas hose, and a pressure nozzle, they drilled away at steep banks of low-grade rock with a stream of water. Though it was a devastation to the land, it was a revolutionary tool for American miners in their search for gold. Water under pressure could cut into rock.

"There is a real pleasure, very distinct, but hard to describe, about the gigantic force [coming from the nozzles]," wrote a reporter for the San Francisco *Bulletin* in 1879, as he watched the water-drilling operation at the North Bloomfield mine in California. "Large boulders and lumps of pipe clay are slowly washed down to the bedrock for the blasters to handle, but rocks two feet in diameter fly like chaff when struck by the stream."

Okay, so Ned didn't invent the thing. But even when I found that out, I still felt proud of our water drill, of the way Ned had looked at the water coming from the pipe, then at the ice thirty feet away, and had seen the drill. I wanted to be able to do that one day—see a situation so creatively that I'd recognize a connection between widely disparate things. If it ever happened, I was sure it wouldn't matter what I made of the connection; the recognition would be triumph enough.

So it was probably a combination of things that lured me up the Ompompanoosuc to its headwaters that summer of 1992: the time I'd spent at Ned's sugarhouse listening to Great Falls during the last couple of years, the occasional walks we'd taken downstream from the falls, and the whole notion of seeing more creatively, which the water drill had introduced me to. When it came to creative seeing, a river seemed like a promising place to experiment. Limited at one end by its head, at the other by its mouth, on the sides by its banks, and determined by a one-way current, it seemed designed for predictability, or more predictability than the old-growth-cleared-regrown-inadvertent woods had shown over the last five years.

Also, there was something about a river's source that especially intrigued me. The place a river started: it sounded magical. I was curious about what it looked like. I had an image of it in my head, though it was, I'm now aware, as idealistic as my image of wilderness had been a couple of years earlier. It had been shaped in large part by two things I'd read: by the brook in Robert Frost's poem "Directive," "Cold as a spring as yet so near its source, / Too lofty and original to rage"; and by Meriwether Lewis's journal entry for August 12, 1805, the day he and William Clark and the rest of the Corps of Discovery reached the headwaters of the Missouri River. He wrote: "The road took us to the most distant fountain of the mighty Missouri in search of which we have spent so many toilsome days and restless nights. . . . Judge, then, of the pleasure I felt in allaying my thirst with this pure and ice-cold water which issues from the base of a low mountain or hill of a gentle ascent for ½ a mile." So, that's what I thought I'd find on the Ompompanoosuc—a remote copse where a single surge of water, clear and smooth, rose out of the ground like a fountain, making a delicate rainbow arch before heading downhill. Wasn't that how all rivers started?

Above Great Falls, I followed the east branch through Post Mills and West Fairlee, and eventually into Vershire, tracking the river north, and then west, as it curved and forked. Thinking that the true source certainly existed on the main stem of the river, I tracked that branch first. As I followed it, I was amazed by the land's transformation between West Fairlee and Vershire. Climbing toward the headwaters in those six miles, the valley changed from a broad, gentle crescent to a narrower, steep-sided V, and the river's meanders became tighter, hugging close to one ridge wall, then cutting sharply across to the other, then cutting back again a little further on.

In Vershire, where the river disappeared on the map, I thought I'd found the headwaters more than once. The first time, it was at a small pond tucked behind a house for sale. The pond was dressed with a white sand beach, and toys lay scattered on the sand. What a fittingly innocent place for the river to start, just how I'd imagined it and so much like Frost's "Directive," which also described a children's playhouse beneath a pine near the headwater brook.

Only one problem: I could still hear the sound of trickling water. Looking to my right, I saw the pond's neatly staired inlet, and then noticed the stream disappearing uphill into the woods. This was not the end of the line.

So Griffy and I followed the river further, about a quarter of a mile, until we came to another pond, this one set behind a commercial garage. The pond was L-shaped and much larger than the first. It was beautiful too. On the south side, directly across from me, a hill rose steeply from the water. Hemlock and birch lined the shore on that side, the birches' reflections striping the water like bleached bones.

I walked around the entire rim, Griffy up ahead of me splashing through creases of wet ground. As much as the first pond, this one seemed a suitable spot for a headwaters, but even more than the first it had a noticeable inlet—a wide, silty cattail marsh I sank knee-deep in as I traced the river upstream. I followed it through some woods, and across a narrow gravel road, and up into the woods again, where it disappeared, just like that. On one side of the road there was a mud puddle, on the other side I was kicking through dry leaves.

"Well, *I'm* inspired," the voice said. "How 'bout you?"

"I don't know."

"Am I to understand that the mudhole on the other side of the road is the source of this river?"

"I don't know," I said. "It's not what I expected."

"Me neither," he said. "You had me all excited."

"Yeah. Me too."

"So, where'd you screw up?"

I sat down. "I don't know."

"Enough with the 'I don't knows'! You're killin' me. Say somethin' else, will you? So I know your brain hasn't cramped up."

"What do you want me to say? I thought I did everything I was supposed to. I followed the main stem of the river as far as it went, and here we are."

"Well, that leaves us with two alternatives," he went on, sliding into the soft, supportive kind of tone I used when I was teaching. "One, what you're looking for doesn't look anything like you think it should, or two, you don't know shit about what you're looking for."

"Maybe it has something to do with the season," I said. It had been a dry summer around here, so the levels of both the water table and the river were lower than normal. Even down near Thetford the Ompompanoosuc was more a stream than a river, using only a part of its rocky bed, leaving the rest exposed like quarry rubble. "And if it were wetter, the mudhole might be a brook, and the brook might go under the road and up the hill there into the woods toward the real source."

"Okaaay," he said slowly, now exaggerating my teacherly tone, taking on a high gentility. "Very good, Terry. That's a possibility. There's no sign of any kind of watercourse up there, but that's a good idea anyway. Now, let's try it again, and this time let's try to remember that we live in the REAL WORLD!"

"Alright, alright, alright, alright," I said, laughing. "I take it you have an explanation."

"Oh, I might be able to offer one, my good man," he said, still genteel. "Though, of course, it isn't appropriate to say in polite company. What's essential, however, is that we leave here at once, because this is a very embarrassing situation, you sitting here all alone next to a mud puddle. Embarrassing to everyone. *You* look quite pathetic, and *I'm* humiliated even to be seen with you. The only thing worse would be if someone else saw you. It would be very hard to explain."

"Okay, sir, we're outa here."

"Thank you."

"But a mud puddle as the source of a river?" I asked, half egging him on. "Deal with it!" he barked.

I was grateful for this kind of banter with the voice, whenever it came. Not the banter itself; I was used to that. Like two people stranded alone on an island, we'd learned that humor was critical to our not killing each other. In the past, though, the voice's humor had always seemed to be serving a more dangerous ulterior motive—a way to soften me up, make me easier to kill later on. So I'd had to be wary, even in lighter moments.

But now there were times when I didn't sense any edge to him; they seemed to happen whenever I first started on a new project or idea, as if he were granting some leeway for my ignorance. In the past he'd always been quiet during those times. The first year on Cobble Hill, for instance, he'd let me fumble along alone for a while; it wasn't until I was wandering around in the swamp trying to find peepers that he'd stepped in angrily. But these days he was around from the start, and during the leeway period he wasn't terrible to be with. Aggravating, maybe. But not dangerous. His dangerous side usually came later.

Looking back, I know now how dramatic the voice's change was, how important it was to my unfolding recognition of my depression. But I didn't really know what to make of it then—whether to be happy he'd lightened up or scared he was around more. I didn't know whether to presume I was feeling better or worse. But regardless, I was grateful for the shift, for the sense of safety it provided now and then. And at some level I also understood why the shift had happened: it had been the landscape's influence. The valley had uncovered the voice's tolerance and drawn it out like moisture.

Sweaty and covered in mud to my thighs, I walked back to the car, where I toweled off Griffy and myself. Then I turned the car around and started back down the gravel road. Suddenly a man stepped out of nowhere. He was a muscular guy with thick hands, sharp eyes, and a handlebar mustache. Flagging me down, he walked up next to the car and through the open window identified himself sternly as the owner of the road I was on, which was his driveway. He wanted to know what I thought I was doing there.

"Good going, slick," the voice said to me. "You've done it this time. Explored yourself into some real trouble. I told you we'd run into somebody, but did you listen? Noooo. You just had to sit there in the mud and give Paul Bunyan here time to find us. Shit, we'll probably end up as his dinner tonight."

In my friendliest, most polite manner I told the man who I was and where I was from. That seemed to soften him a little. I think he was worried that I was casing his house.

The voice piped in again. "Wait. Don't kowtow to this asshole. Now that I see him up close, I'm pretty sure you can take him. Just tell him to shut up and mind his own business, then get out of the car and kick his ass."

"Yeah, right," I answered. "*You* shut up. You're going to get me killed. This *is* his business. I'm on his land."

"And a prime piece it is. Tell him how much you love the mudholes."

Instead, I told him why I was there, and when he heard that and sized me up—saw my sweat-soaked shirt, my wet pants and boots, and the muddy dog in the seat beside me—he relaxed even more. He could tell I was in no shape to burglarize anybody.

He introduced himself as Jim and then pointed behind him to the L-shaped pond. "That's the headwaters right there," he said. He'd dug it out himself, excavated it with a bulldozer years earlier, when he'd worked for a wilderness camp located there. After the camp closed, he'd bought the pond and the land around it, then built his house and the commercial garage.

I asked him if there might not be another spot where the river started, someplace further on. I said I'd followed a stream into the woods above the pond, and it looked like it went somewhere in wetter times.

It didn't sound like I'd followed the main channel up from the inlet, he said, because the main channel went up to one other pond uphill from his house, a smaller pond belonging to a neighbor. It was a pretty place, and it did drain down here, but he didn't think it was any more the headwaters than this pond, and he hadn't been back there much because the neighbor didn't like people trespassing.

I couldn't tell whether he'd meant the "trespassing" bit as a hint to me or not, but I played it safe and took it as one. And so, after thanking him for the information and apologizing for walking across his land, I drove off.

I never did go to that last pond, even though I had the sense that it *was* the last pond, or the first, the starting point of the river. You couldn't go much further in that direction before coming to the height of land anyway; beyond there everything sloped west toward the town of Chelsea, and was part of a different drainage, the White River watershed. So topography almost ensured that it was the first pond. But still I didn't go.

My reasons were pretty unimpressive. There wasn't the intuitive understanding I'd felt in the secluded field I'd come across five years before as I was snowshoeing toward home across Cobble Hill; and there certainly wasn't the nobility of W. D. Wetherell's decision in his book *Vermont River*, when he traces his favorite river in the state toward its source but deliberately turns back before getting there. He writes, "There are mountain summits in the Himalayas that are sacred to the peoples living at their base; climbers, respecting this, stop a foot short of the summit, leaving that final temple of snow forever virgin. It was out of this same worshipful impulse that I turned back from the source."

The main reason I didn't go was this: I'd just been caught trespassing. It may not have been a big deal, but I hated it. Hated the hand-in-the-cookie-jar exposure and the potential confrontation that made me feel guilty and angrily defensive at the same time, my heart racing all the while. Jim's appearance had stirred me up, and now I needed some time to calm down.

But there was a lesser reason too, something about the whole headwaters idea: I could feel it changing. When Jim had told me about the last pond, had said it wasn't any more the headwaters than his own pond, it had sounded at first like bravado—posturing words from someone who knew he owned the second-best version of something. But the more I thought about it, the more I saw his point. I imagined myself going back there and standing on the pond's bank, assuming I was looking at the definitive source of the river, yet still not being sure. How could you ever be sure? And if you couldn't be sure, how could there be such a thing as a single, identifiable source?

Those questions had been building all day, even before I'd reached the muddy patch below Jim's driveway; but they'd grown louder as I walked around the L of his pond. Two or three times on the south bank, traversing the shoreside woods, I'd noticed Griffy crossing a trough of wet ground; when I got there myself and my eyes followed the line of wetness uphill, I could see the hillside draining into the pond in each of those places. As far as I could tell, there wasn't just one source of water. There were many.

So, driving home that day, I was readier to consider a different kind of headwaters. Perhaps somewhere in the Vershire hills there *was* a Lewis-and-Clark source—a rich, secluded, ridgetop spring spilling into a channel that became the Ompompanoosuc. If so, I hoped to come across it some day, if only to revel in its beauty. But as far as its being the river's single origin, that didn't seem so realistic anymore. What seemed likelier was that the hills were perforated like a sprinkler head, that a river's source might in fact be everywhere.

The next day I tested the idea while tracing a different branch of the river toward South Vershire. Part of the drive took me through what had been the town of Ely, a mining village that sprang up around the copper mine there and flourished toward the end of the nineteenth century. Describing the town in his book *Green Mountain Copper*, Collamer Abbott writes: "This site contained more than 50 houses with perhaps 100 tenements, a giant smelting shed, a large store, and two churches, plus all the installations required to conduct a mining and smelting industry that reached peak production of more than 3,000,000 pounds of copper in 1880."

Now, more than a century later, no visible evidence of it remained. At least none I could see from my car. Had I wandered around on foot I might've run into holes in the ground that had been house cellars, or glimpsed some rocks in the river that had been stained orange by copper tailings. But driving by that day, the village was essentially invisible to me. All that struck me was what beautiful country it was around there— the hillsides thickly wooded, the valley alternately wide and narrow, the road winding beside the river, the river disappearing deep into a ravine as the terrain climbed to a plateau at South Vershire.

At a T in the road the river seemed to curve left behind a house. So I stopped there and talked to a nice man named George, who told me that what I thought was the river was really the runoff from a gushing artesian well some people had drilled for their home on the hillside. Just after the well was drilled, he said, so much water started draining downhill from it that a neighbor lower on the slope dug a pond, diverted the runoff into it and filled it up. He pointed to where it was.

The stream I was looking for, though, went off to the right and then up the hill. If I wanted to follow it, he said I could either walk up a nearby road that cut through some neighbors' fields, or drive around to the top of the hill and start down from there.

I thought the nearby road sounded best, so George told me who owned the fields and directed me toward their house. I wasn't going to take any chances today: if I was going to walk across somebody's property, I wanted permission first.

Standing outside their screen door, I introduced myself to the woman on the inside, who kept the door between us as I explained that I was tracking the river and then asked if I could hike up her hill.

She looked at me quizzically. She must have been wondering what my real angle was. Usually people wanted to know if I meant to hunt on their land, especially when they saw Griffy, who looked like a coon dog; few ever seemed convinced that my motive was curiosity alone, and I couldn't blame them. I mean, how many people went around tracking a small branch of a river just to find where it started?

But I must have looked innocent enough standing there, because finally she said, "Sure, go on up," and nodded toward the road behind me.

It was a grassy farm road heading straight uphill, bordered by maple and beech, hayfields lying beyond the trees on each side. I headed uphill with it. In the woods above the fields the road narrowed to a path, then gradually thinned and finally disappeared, leaving me to bushwhack on my own. While Griffy scampered off, chasing chipmunks, I turned south across the hill, picking my way through young maples toward where I thought the river would be. It was hard going and I couldn't find any water at all, let alone the river, so I retraced my steps back down into the field, where I could cross the hill more easily. Turning south again I walked to the edge of the field and then just a few paces into the woods. There it was, the stream. I turned uphill and followed it.

Eventually it slowed and narrowed and dried and finally vanished, in an uneven progression I recognized from the day before. There was no discernible origin to this branch, no single point where I could have stopped and said, "This is where the stream begins." Instead, there was watery ground, and then before I knew it, I was wandering across a dry hillside. I traipsed back and forth through the high woods, unable to locate any further sign of moisture.

So I worked back down the slope toward the last place I remembered seeing wetness; after a while I stumbled upon a patch of damp ground with an oval puddle in it. Griffy was as happy to see the water as I was; he drank from it, then lay down in it to cool his ribs and belly.

I walked slowly down, noticing how the wet patch gave way to dry ground, then another wet patch, then dry ground again. Further downhill I heard a clacking sound, like tiny castanets, the delicate percussion of running water. I was off. Weaving among maple and hemlock, I found the brook in a narrow crevice and followed it down, jumping and skipping after it as it widened and flattened, watching one brook join it, then another and another, four or five in all, each coming in unexpectedly to build this hybrid flow and make it run faster.

At one of the lower forks I stopped. Turning, I looked back uphill into the woods. I'd been so excited about finding the moving stream, so swept by its quickening current, it hadn't occurred to me until then that I'd just witnessed the birth of a river. Or *a* birth of it. And it was strange: even though I'd located that isolated puddle up there, its water still and clear, and then watched it grow gradually into this quick, splashing stream, I'd have been hard-pressed to explain what had happened, or how—how the river had happened. All I knew was that it had, and that it had rolled me along with it.

And I was reminded of a time a decade earlier when Robin and I had climbed a Sri Lankan mountain called Adam's Peak. At least that's what it was called in English. Known to Buddhists as Sri Pada, the "sacred footprint," and to Hindus as Shivani Pattan, "the footprint of Shiva," the 7,000-foot mountain was a holy spot, a goal for pilgrims, who journeyed to the summit for two reasons: to see the rock concavity said to be the footprint of Buddha, or Shiva, or Adam, or St. Thomas, depending upon the slant of your faith, and to watch the miraculous sunrise, when the mountaintop cast a perfectly triangular shadow to the west.

I was reluctant to make the hike, particularly at 2 A.M., the hour we'd have to start in order to reach the top by sunrise. It all sounded like an elaborate tourist trap to me. But Robin had the breath of adventure and faith in her. Maybe a little naïveté too. She convinced me to go.

Intermittently during the fifteen years we'd been together, especially after rough spells, we'd wondered aloud to each other what it was that had kept us together, so mismatched a couple we seemed: Robin the organized, gregarious optimist; I the pessimist, always intense, at turns brooding and brash. She the reconciler who'd become a psychotherapist; I the loner, protective of my creative space. The best we could figure was that there was a fatedness to our love, almost a necessity; it was as if we balanced each other, and inside we knew enough to hang on to that.

As hard as it was for me to admit, I recognized it was true. Sometimes Rob could get me to do things I normally wouldn't, believe in possibilities that to my mind seemed unrealistic or naïve or that courted failure. The climb up Adam's Peak had been one of the earlier instances; conceiving a second child was the latest.

"What if our next child is just as hard for us?" I'd asked.

"It won't be," she said. "Look at our friends. Their second children have been so much easier for them."

"But what if ours isn't?"

"Ter, it will be."

"You can't promise me that," I'd said.

"Of course I can't."

"So, what if ours isn't?"

"Then we'll be better prepared, won't we? We will have been through it once."

"I don't think that's going to make a difference to me, Rob. I still am who I am. I only have so much patience."

"But think how much we've learned about ourselves as parents. You didn't know your limits when Carry was born. Now you do."

"I just don't want to feel that out of control ever again. It scares me to remember some of those times."

"I know. And I'm sorry for that, sweetie. I felt that way about myself too. But just look at how much things have changed. Look at how close you and Carry are now. He loves you so much."

On and on the discussions went, for months, with Robin's hopefulness finally softening my fear, coaxing me to look behind it, helping me remember that I'd always imagined having two children, not one, and that I did want a second chance to be the kind of parent to an infant I'd imagined I'd be. So, while a part of me was terrified of what might happen, we'd gone ahead and conceived.

In Sri Lanka ten years earlier I hadn't felt fear so much as skepticism. I didn't want to climb Adam's Peak at all, but when Rob started up the trail that late-January night, near the start of the pilgrimage season, I was beside her. The trail was actually a well-lit staircase; there were tea-houses every so often, where you could stop and rest and have something to drink or eat. Near the top, the path steepened and we pulled ourselves up the last stairs as if climbing a ladder. There were probably a hundred people already there at the summit. It was sometime after 4 A.M. And it was cold.

Our first stop was, of course, the footprint. Unfortunately, it was what I'd feared it might be: a large shapeless dent in the rock, one I might have been able to chip out myself with a pickaxe. (I've since read in a guidebook that what I saw *was* a large dent, deliberately carved to look like the original footprint, which supposedly lay protected and encased beneath the visible "footprint.") I couldn't fathom what might have led that first person to imagine a footprint there, except exhaustion from a rugged climb or exultation at reaching the summit. Or a rabid faith.

All we could do was wait. We tried to lie down and rest in a shelter, but it was too cold to be comfortable, and the stone floor only magnified the chill. The other visitors there—many of them pilgrims, clad in white—couldn't have been comfortable either. They were dressed even less warmly than we were, and sat against the shelter walls, closely packed together. That's what we did too. When the sky brightened two hours later we made our way outside and stood facing east on the broad cement grandstand.

By then I was beyond being skeptical; I was annoyed. Cold and tired from the climb, sleepless, betrayed by the footprint, I figured a fitting end to the experience would be a cloud-choked dawn. But when the sun came up bright, when it climbed and exposed the fog-covered land to the east, I was humbled. Looking east, I could see just how airborne we were; it seemed as if we were being dangled from the sky like marionettes. And when I walked to the other side of the summit and saw the triangular shadow on the mists to the west—*my god, a perfect equilateral triangle; how does a cragged mountain make a straight-sided shadow like that?*—I knew at last why someone would want to make this spot sacred, would see a footprint in the hollow of a rock: something would need to justify the miracle of that dawn.

Balanced on a small, pointed headland in South Vershire, flanked by two lines of tumbling water that had been three or four brooks higher up, and who-knows-how-many puddles above that, I had the same feeling. Up in the quiet woods there had been smooth, leaf-covered ground with a couple of shallow puddles scattered about; down here there was a noisy, branched stream. And even though I'd witnessed it first-hand, I couldn't have told you where or how the smoothness had turned creviced, how the quiet had become noisy, how the dry was made wet. At least not in any clear, exact way; there was too much to include, more than I had the words for. Yet it had seemed so simple at the time.

And it was then that I remembered the aura of Adam's Peak, because in my befuddlement about this river I imagined I was feeling the same need the first climbers of that mountain must have felt—the need to explain an overwhelming natural phenomenon with something definite, precise. They'd found a sacred footprint to answer that dawn, and all of a sudden what came to me was an image of the only thing mystical enough to answer this river's transformation: a single source of water.

The rest of the walk was a frolic. I wandered downhill beside the stream and then in it and sometimes across it on fallen logs. Occasionally I followed a thin path that wove back and forth through the water. When it crossed the muddy flats, I could see from the pointed footprints that it was a deer path.

Like all the rivers that summer, this one was low, carrying only a fraction of its capacity. That left most of its courses exposed; it was easy to see the maze of bows and braids the water followed when the river was full, when it swelled out of the main bank and cut other detours downhill. Now those detours were just empty dirt troughs; I saw in their beveled shape the kind of cuts I used to make in elementary school art class when I carved pictures into linoleum blocks with a gouge.

Those gouged lines must have been lingering in my mind when I visited Great Falls a couple of days later. Twenty-five feet wide at their crest, dropping forty-five vertical feet, the falls are gorgeous, reputedly the largest undammed falls in an open landscape in Vermont.

They hadn't always been undammed. As Charles Hughes describes in *The Mills and Villages of Thetford, Vermont*, there was a long history of mills at the site. By 1815, fifty years after John Chamberlain had arrived in East Thetford, there were certainly three dams on the falls at once, one coming after another, descending like steps, each supporting a mill: a gristmill, a cloth-making mill, and a sawmill. A hundred years later, after the mills had changed hands and uses many times, there was only one left, owned by Charles Vaughan, who eventually redesigned it so that it would generate electricity—the first electricity in Thetford.

Now the falls ran freely. No mill or dam stood on them, only remnants, a condition that seemed unlikely to change, since they'd recently been protected under a special designation by the State of Vermont.

Over the years I'd found myself watching the falls and listening to them whenever I could, smelling and touching and imagining them from as many vantages as possible: from the safety of the border rocks in spring and fall, while water burst downhill like horses through a narrow pasture gate; from the center rocks during the August lull when the river trickled gently over the old, broken dam wall like thin braids of hair, when it was possible to hop the uneven rock terraces down the middle of the falls without getting wet.

That's how low the falls were when I visited them this time; I walked out onto rocks normally deep under cascading water. Sitting down, facing downstream, I noticed something strange: small pockets on the faces of the rocks. I leaned over to look at them closely. Oases on dry land, these pockets were. Little rainwater pools. There was no pattern to them; they had different shapes and depths. But they weren't metamorphic deformities, as far as I could tell. The edges and corners were too smooth, the cavities too carefully shaped. It was their placement that finally gave them away: each sat below a small channel mouth, like the indented landing spot at the end of a playground slide. All at once I knew they'd been pocked by the repetitive slap of water.

In E. L. Doctorow's novel *Billy Bathgate*, Billy, a quick-thinking street kid from the Bronx, finds himself out of his element in the Adirondacks. As he looks at a waterfall, he says, "It came home to me that falling water is what makes gorges, I mean this could not have been news to anyone, but it was practically the first bit of nature I had seen in operation."

Sitting on Great Falls that day, it was as if I'd just stumbled out of the Bronx myself. I looked up from those dimpled rocks and recognized for the first time that the land all around me was being drilled slowly and relentlessly by water.

"My god. Look at that," I said.

"Took you a while," the voice said. "And to think it was staring you in the face the whole time."

"Oh, you knew?"

"Of course."

"Of course."

"What'd you expect?" he asked.

"Not this," I said. "Not any of this." I gazed up at the surrounding hills again. "It's amazing, isn't it?"

"I guess it is," he said.

Recognition is such a curious thing. It seems to happen in a flash: a revelation without any warning, a shooting star in a quiet sky. And yet in retrospect it's so easy to look back and see the signals, to say, "Oh, and this is how it must have started, and this led to that, and that to the next, and then I went here and . . ." In this case I could easily see the path this recognition had taken during the last several months—through the one-and-a-half-inch plastic pipe and the Matteson-Miller-Chabot canvas hose and the culvert-steaming rig. Through the brook's crease in Ned's upper fields and the looping gouges in the stream in South Vershire and this swath of waterfall at the valley floor. But what about the less visible paths, further back, that seemed part of this same course too: my interactions with Carry, the legacy of our sugarhouse, Adam's Peak, Robin's optimism, my elementary school art class? If they were all as related as they felt, then this recognition, like the water that had inspired it, had been building slowly and relentlessly for a long, long time.

That we live on the contours of a water-drilled landscape probably isn't news to anyone, though it was to me that day. That we've adopted the water-drilling process and have devoted ourselves to improving it for our own immediate needs is a predictable human twist. But that we unknowingly replicate the process, mimic the river's eroding course with our own thinking, with ideas that spring undetected from remote ground inside us and ride tendril trails toward awareness—to me, that's the most compelling notion of all.

After that moment on Great Falls, it was impossible to look at the valley and its undulations without thinking of water. A sheet of frozen water had scraped down the landscape, and a river had carved it. Even before the last glacier plowed south through here about eighty thousand years ago, all of the rivers had been at work. This one we call the Ompompanoosuc, seeping from the hills to the north and west, had been cutting across twenty-four miles of uneven ground, dropping fifteen hundred feet to today's Connecticut, which was itself chiseling through four hundred miles of south-tipping terrain while dropping twenty-five hundred feet along the way, carrying upland tailings to the sea.

It was also impossible to ignore the river's cue, its rhythm and patient work. How could I? I'd been throttling the terrain to make it cough up knowledge these past five years; now I saw that it would be wiser to loosen my grip and let whatever knowledge was going to come, come in its own time. That was easier said than done, of course. Patience wasn't natural for me. I'd always chased fast, concrete results in order to muzzle the voice, a habit that had seeped into the rest of my life. It didn't matter if it was with the landscape or Carry or Griffy, or with Robin, now pregnant again; I'd been afraid to allow things or people or relationships to progress slowly, especially if it gave the voice a chance to make me feel stupid or uncomfortable. But now I could see that they'd been progressing slowly anyway, and changing too, with or without my conscious help. Even the voice had changed.

When I finally saw the shape of patience in the valley, I began to feel for the first time its irresistible draw within myself. It held a depth and honesty and a freeing quiet that was so different from the combative surface noise of my everyday thoughts. I wanted to be filled with all of that. So I vowed to try, as often as I could, to be faithful to the pace of the landscape, to let my life and self be etched by the current purling around me, and trust the slow, meandering way it brought things to mind.

Chapter 5

The Face

I should have seen the unsettled weather coming. I should have seen it in the south wind. The wind didn't often come from the south; that was the whole point. It usually blew out of the north or west. But whenever it swung around, dragging heavy air up from the mid-Atlantic coast or the Gulf of Mexico, it dragged things out of their normal order too, flicked the months ahead like abacus beads. In January, it could feel like April; this day in May, it might as well have been July. I should have known it from the heat.

During the morning I was at home, writing in my office upstairs, while Robin was downstairs with Carry and Jacob, our second child, now eight months old. I had to give Rob credit: she'd predicted it. The second child had been easier. Much easier. But little of that ease, we knew, had to do with our parenting; it was mostly due to Jacob himself, his temperament. He'd entered the world so differently from the way Carry had: he was calm, happy, fully settled. None of infancy's trials ruffled him. He just went along patiently with whatever the family program was, his eyes bright, a smile almost always on his face. It was as if he'd done all of this before, been around the block a few times. Sometimes Rob and I felt like we were in the presence of a Buddhist monk whose life wisdom dwarfed our own. Among the things we called him was "the old soul."

The contrast between Carry and Jacob had us shaking our heads. How two children conceived by the same two people and carried in the same womb could have such distinct temperaments, we could not explain. But the contrast humbled us, because it revealed how small a part our parenting would play in forming Carry's and Jacob's core personalities. That development, it was clear to us, had happened in utero between conception and birth. Our job now as parents was to be their stewards—not to *make* them into who they would be, but to help them discover and become the best of who they already were. It seemed no less daunting a job.

Sometime in the middle of the morning Robin's friend Laura came over to visit; while she stayed inside with Rob and Jacob, her two kids— Wendy, who was seven, and Will, five—went outside to play with Carry. Given that there were four kids under the age of seven in and around the house, it was surprisingly quiet. Not that I hadn't learned how to shut out family noises whenever I was working at home, but it was still easier when it was quiet and I didn't have to worry about it.

Unfortunately, the quiet didn't help my writing at all. By the time I took a break, a little before noon, I was down on myself, frustrated with what had come out. The piece I was working on had to do with a pattern I'd seen in the local landscape—"bookended wilderness," I called it. There was an episode I'd already written that seemed promising, about a dawn drive I'd taken with Carry along some back roads of Thetford. Now I was trying to follow it up, expose the landscape pattern and explain what I'd learned about it as I'd researched it. Nothing I wrote seemed to work. It was all heavy and pedantic. Plodding. And no matter how many times I reworked the sentences, I still hated them. Hated them more, in fact, because they'd taken more effort to produce but sounded just as flat. So when I looked up at the clock and saw that the morning was almost gone and then looked down at what little I had to show for it, my spirits sank. In their place came that dreaded welling of insecurity.

Determined to escape the desk, I headed down for a snack. It wasn't until I reached the bottom of the stairs and heard murmuring in the next room that it occurred to me I might see someone. From upstairs the house had been quiet enough to be empty. Now that I was downstairs, I hoped it actually was: I was in no mood to interact.

But when I turned into the dining room, there they were, Robin and Laura, talking quietly at the table, their heads bent close to each other.

I said hi to them as quickly as I could, then tried to slip into the kitchen. Looking up, Rob said, "We've had a little problem." She seemed troubled.

"What?" I answered.

"We just found out that Carry took his bow and arrows out without telling us."

Though he was just about to turn five, Carry had shown a love for archery for two years already. From the start he'd persevered through months of frustrating failures, as he learned to nock the arrow, balance it on the bow's arrow rest, draw the bowstring while keeping the arrow nocked, and let go without having the arrow drop to the ground in front of him. He'd shown a patience we hadn't seen with any other activity, bending down over and over and over and over at first to pick up the little rubber-tipped arrow at his feet. After a time he figured out the technique and became surprisingly skilled at it. Surprisingly responsible, too, about its danger as a weapon. So in the last year we'd let him change from rubber tips to target arrows to use with a larger toy bow we bought. But we made sure he understood that he was only to shoot at targets, only when one of us was supervising, and never when friends were over.

So he'd just broken all three of our major rules.

"You're kidding me," I said. "He did?"

That wasn't like him.

Rob went on, "And when Wendy got a turn to shoot it, she pulled the string back and pointed it at Will." She was quiet for a minute; I waited for the gory result. "Will came in and told us. He was really upset."

"He wasn't hurt?"

"No. Just scared."

"I don't blame him," I said.

I didn't know what to think, didn't know if the two of them expected me to do something or not. What was I supposed to do? On one hand it seemed like one of those near-misses of childhood you chalked up to luck. I mean, kids broke rules, right? We all did when we were kids. That was part of being a kid, wasn't it? All you could do as a parent in those cases was pray things would come out all right, and when they did, as they had this morning, sigh in relief and not think too hard about what could have happened.

But I'd just spent the morning banging up against my deficiencies as a writer, so I wasn't feeling particularly philosophical.

"He really took out his bow and arrows?" I asked Robin again, my voice thin, incredulous.

She shrugged her shoulders and nodded.

"That's just not okay at all," I said, anger slipping into my voice. "Didn't we make it clear to him?"

"I thought so," she said.

I shook my head. Maybe it was the July heat in May, or the serious-ness of Robin's and Laura's expressions, or the hours of failed writing upstairs, or the insecurity spawned by the failure, or the anger now growing out of the insecurity. Or all of them together. But suddenly, as if I were a balloon attached to a running faucet, emotions were filling me up, swirling around inside, making obvious what I'd been fearing all morning.

"It's all your fault," the voice said. "That's what they're tellin' you."

"Hey, I don't need that from you, asshole. Shut it," I said.

"Ooooo, testy. What happened to shape of patience in the valley and the vow you made to honor it?"

"It's gone. Now, shut . . . up," I said again, but we both knew he was right. Standing there in front of the two mothers, I felt instantly respon-sible, as if I'd handed the nocked arrow to Wendy myself and watched it sling accidentally from the flexed bow in her hands, flying into Will's eye or throat or wherever. As I looked at Laura, I imagined I saw in her eyes a blame that intensified my own, and suddenly I could sense dark-ness coming on inside me and the debilitating weight and fatigue that came along with it.

All at once the air in the house seemed poisoned, full of my own wrongdoing. "I gotta go," I lied. "I guess we'll talk about this when I get back."

"Where are you going?" Rob asked.

"Uh, I gotta do some fieldwork," I answered quickly. What I meant was, *I gotta get outa here.* Grabbing my backpack, I dropped in my field guides and binoculars and threw in a bottle of water and some fruit. "See you later."

I was gone.

Sometimes the heaviness came so fast that it was like being caught in rocketing elevator and not being able to free myself from the floor, because the gravity was too strong, because the floor was already there, above where I'd have been had I sprung upward. Suddenly I'd feel tired and pressed flat, like glued linoleum.

Anything could bring it on: guilt from losing my patience with Carry; envy at a student's best work; self-doubt when an editor's rejection arrived in the mail; annoyance with the babysitter after her car had broken down, forcing me to cancel class at the last minute and play house husband. It was never clear to me why it chose one event over another, but it didn't matter. On it came anyway.

When I felt this way, I usually obeyed its physics and lay down, with no plans to get back up. Some mornings I couldn't understand why my eyes were open, couldn't imagine being grateful for anything. Lifting my torso from the mattress or swinging my legs over the side of the bed seemed not only impossible but pointless; it meant confronting a dimness that was another day further from nothing and one day closer to my death.

But today, there was no lying down. I had to get away—away from Carry's bad decision and the image of Will's wounded body and Laura's blaming eyes, away from my own sense of final responsibility. Or I had to feel like I was getting away. There was no real getting away. I knew the darkness and the voice would come with me. But I also knew I'd have a better chance against them if I was alone in a place that didn't seem to judge me. That must have been why I'd mentioned "fieldwork" without thinking. That must have been why I found myself driving off into the valley.

The valley would take care of me.

"What kinda parent are you?" the voice asked once we were in the car. "Your son doesn't even follow your rules. You almost killed that other kid, you know."

"Listen, we both know how I'm feeling. So just give it a rest while I work it out."

"Work it out? This isn't nuclear physics, Einstein. I can work it out for you right now. What happened is that your kid has no respect for you."

"That's bullshit," I said. "Carry's a great kid. Don't say anything bad about him."

"Take it easy. I'm not talkin' about him. I'm talkin' about you. You're a loser. You think you're a writer but you can't write; you think you're a good father, but when somethin' goes wrong you run away and hide in the trees."

"It's not my fault!"

"Yes, it is. You know it is. That's why you're in this car runnin' away."

"I mean it's not my fault I feel this way. I don't have any control over you. I never know when you're going to show up and get in my face."

"You see? There you go again. The bow-and-arrow deal is your fault, and you know it, but now you're sayin' it's not your fault that you know it's your fault. How fucked up is that?"

"About as fucked up as you not understanding, especially when you're in the middle of it all, helping it along."

"Yeah, and if *I* don't understand, how do you expect Carry to? He already sees what a head case you are. It won't be long before Jacob will too. So why should they respect you? They shouldn't. No straight-thinkin' kids would. And that's why this all has happened. It's like they say: when a kid does somethin' wrong, don't look to him. Look to the parents. Then you'll understand why the kid did what he did."

Fifteen minutes later, Griffy and I were in Ely, climbing uphill, the wind blowing hot and thick at our backs.

To get where I was, I'd wandered around on Route 244 for a while before finding someone who didn't mind me walking on his land. His name was Craig and he nicely let me park my car on his lawn and said I could hike wherever I wanted.

"See? *He* doesn't think I'm a loser," I told the voice.

"That's 'cause he doesn't know you," he said. "Tell him about this morning and see what happens."

My plan was to head up the hill behind Craig's house—a hill that didn't seem to have a name—and follow it north toward Lake Morey before circling back somehow. I'd never been on this part of the ridge before, so I didn't know exactly where I was going or how the trip would work itself out. But given what had happened at home, I didn't care.

That I'd chosen this section of ridge to hike might sound haphazard, but it wasn't. Once I'd pointed the car into the valley, I knew where I was headed. I'd been working my way north along the Vermont piedmont during the past weeks and this was the next hill in line. It was also the only piece of the valley rim I hadn't explored yet. So, seven years after first seeing the view, I was finally going to complete the perimeter of it. I liked the sound of that.

When I first began exploring the valley, I organized my journeys quirkily, breaking the terrain into haphazard pieces. One day I'd walk Potato Hill from Five Corners. The week after, the train tracks from North Thetford to Ely. Then I'd canoe the river from Ely to Fairlee. There was no real science to this; each piece was determined only by the ground I could cover in a day or less, and by a commitment never to come upon the same place in the same way twice. Whenever I revisited a spot, I'd alter the direction of approach, or my mode of travel. I'd wait for a different time of day or season. Originally my goal was simple familiarity; whether at Great Falls or on Cobble Hill, I just wanted to feel more at home in the place. I thought the varied perspectives would help, and they did.

But over time I noticed that something different had been happening too: I'd begun superimposing on the valley's contours my own pattern, a private geometry. On top of that oval bowl, I'd reimagined the valley as a semicircle, like a wagon wheel split in half. Cobble Hill was at the hub and the curving valley ridges were the outer rim. And I'd projected my journeys into the landscape like the spokes cast out from home—sometimes all the way to the outer rim, sometimes to nearer spots. Eventually I found myself connecting the spokes with bending lines of travel, which made the whole pattern look woven, divided into compartments, like a spider web.

Though I hadn't planned it that way, I wasn't exactly surprised to look back and see that I'd done it. From books I'd read, I knew that many people responded to new experiences by organizing them rationally, creating some comfort in their strangeness. Isak Dinesen wrote in *Out of Africa*:

> When I flew in Africa, and became familiar with the appearance of my farm from the air, I was filled with admiration for my coffee plantation, that lay quite bright green in the grey-green land and I realized how keenly the human mind yearns for geometrical figures.

Still, this predetermined pattern didn't sit right with me. Something felt misguided, or wrong, about thinking geometrically, especially in light of the vow I'd made to honor the landscape's own pace and shape. What if my spider-web pattern wasn't true to the valley at all? What if all that I was seeing out there was simply my own creation—my reflection in a topographical mirror—and not the valley itself? What a waste of time these years would have been.

The initial ascent up the ridge was steep; the terrain shifted back and forth between rocky clearings and shady hardwood stands. Around my head black flies swarmed like a sphere of crazed electrons; all of their movement made them seem twice as dense as they really were. If I let them alone, they whirred around, thick as risen dust; but whenever I grabbed at them, I came up empty. Nothing in my hand. Then there they were again, thick around my face.

In the clearings were clusters of white violets, in the groves a beautiful mixture of hardwoods—beech, hop hornbeam, ash, white birch. Further on, a surprise: a lone wild columbine. I saw its color from thirty feet away. When someone first pointed the flower out to me years before, I'd instantly fallen in love with it, its redness and crowned shape. Called "columbine," from the Latin *columba*, "dove," its five upward-pointing petal tubes must have looked to some colonists like a circle of perched doves. On the other hand, it got its genus name—*aquilegia*, "eaglelike" —from its petals' resemblance to an upturned eagle talon. Some people have seen the horns of the cuckold in the flower, others a Congregational church steeple. Still others have seen a bell, which is the form I saw most often. But in other moods I'd seen a jester's hat, a drooping chandelier, an upside-down five-legged fountain. With enough time I suspected I could see anything in it. Somehow that seemed part of my problem.

Uphill past the columbine, I glimpsed another splash of bright color in the leaves ahead, too low to the ground to be a flower. I tried to guess what it might be: a glove, a wrapper, some surveyor's tape. When I reached it, I found an oval piece of material—fleecy, fluorescent orange, with two half-moon shapes cut out of the right edge, the top moon bigger than the bottom. Though I could tell what it was right away, the form still took me a little off guard. It was a hunter's hood, flattened to a right-facing profile; it had a haunting, dismembered quality about it.

It also reminded me that no blushing life in these woods came without a predatory shadow.

Near the crest of the ridge, I emerged into a clearing open to the south and west, and took off my pack to rest and let the hot wind chill and dry my soaked back. Tired as I was, it was a good tired, not a dark, heavy one. The climb had left the rest of my life far behind.

In the near distance to the west lay Lake Fairlee; beyond it ran the ridges of the Ompompanoosuc valley, bending northwest toward the headwaters in Vershire. I'd been there. To the south were places down the Vermont ridgeline I'd already explored too—Houghton Hill, High Peak, and closest, Ely Mountain. It had been on Ely Mountain, some weeks earlier, that I'd noticed how deeply my own geometry had influenced my walks. The first day, I'd climbed the hill from the south, then wandered the summit, identifying landmarks. The next day I approached from the north and tried to find the same spots I'd seen the day before. After bumping into a few, I finally chose one to be the hub of my explorations. The third day, I took a different approach from the north, walked a circumference around the summit, a rim around the hub, then zigged back and forth across the top to get a feel for the cross-terrain. Only when I could find the hub's location from different angles could I move on; only then did I feel comfortable there. Now I wondered whether I'd suddenly feel that same comfort today when I finished the circumference of the valley rim.

It had been from the northern slope of Ely Mountain that I'd seen clearings on this hill, the clearings I just climbed through. I had gawked at them, square patches of steep meadow amid the wooded hillside, and promised myself to walk up to them at my next chance. Part of their appeal was the same kind of beautiful remoteness I'd felt about a couple of other places in the valley, gemlike spots that would have stayed hidden from me had I not stumbled upon them or noticed them from an unusual vantage point.

When I did some research about the ridge in the following weeks, I became even more curious about this hillside. The deep gap between Ely Mountain and this peak, the gap that made a view possible, had been called the Ely Wind Gap. It had an interesting history. In his 1950 geological report on this section of the valley, Jarvis Hadley explained:

A remarkable feature of erosion during the period of canyon cutting in the Bradford-Thetford area was the natural diversion of a former tributary of the Connecticut River near Ely, Vt., leaving a wind gap now occupied by the road from Ely to Lake Fairlee. This gap, 700 feet deep, is the only large gap in the western wall of the Connecticut River valley between the Waits River and the Ompompanoosuc River. It is particularly striking as seen from the upper slopes of the 1,500-foot hill north of the road. No stream flows through this gap, and it is clear that the stream that once carved it has disappeared. Moreover, a study of the drainage . . . strongly suggests that the North Branch of the Ompompanoosuc River . . . formerly flowed eastward through the gap as an independent tributary of the Connecticut River. Some time early in the period of canyon cutting, a smaller tributary of the West Branch of the Ompompanoosuc, working rapidly northward, intersected this independent stream, which we may call the Ely River, near Post Mills, thus capturing it and diverting its water into the Ompompanoosuc drainage system.

After thousands of years of cutting through the ridge, the Ely River had changed its course, joined the Ompompanoosuc, and headed down the west side of the ridge, where the Ompompanoosuc had already done its work. At first I regretted the course change a little. It seemed like a lot of wasted energy on the Ely River's part; having done all of that work, the river ought to have been able to enjoy it.

But then I had to stop, remind myself that I was making a value judgment similar to the one Hadley had made later in his report, when he called the change of course an "act of piracy" on the part of the Ompompanoosuc. His tone was tongue-in-cheek to be sure, but not insincere; I suspected he was simply overdramatizing the common term for that kind of diversion, which was "stream capture." And I didn't want to get caught in that same assumption.

Had humans consciously diverted the river for their own advantage, it certainly could have been judged a kind of stealing. But as it was, what had happened was the natural movement of rivers, the same thing Leonardo da Vinci had noticed five hundred years ago: "The windings which the rivers make through their valleys as they leap back from one mountain to another cause the bank to form curves," he wrote in his notebooks, "and these curves move with the current of the water and in course of time seek out the whole valley."

In this case two tributaries seeking out the same valley, winding where the terrain allowed, had intersected and then followed the less precipitous course to the main stem of the river. The landscape had reorganized itself, redefined its own shape, and in the process left this gap—the lasting afterimage of eastbound water. But there had been no crime involved; the morality we applied to it was of our own making.

Looking just east of Ely Mountain now, I pointed my binoculars toward home. It wasn't a gesture of conciliation for what had happened earlier; it was habit. Searching for our house was something I tried to do as often as possible on my walks. For the most part I'd been unsuccessful; the valley's wooded terrain rarely allowed the right line of sight to Cobble Hill. Only from the Palisades, where the unobstructed view was straight back down the valley, had I had any luck.

The whole idea was hokey, I knew, but I couldn't resist. There was something exciting about the chance of seeing our house from far away —boyishly exciting, as if I were in elementary school, trying to draw fawning attention to myself to fill the dark emptiness inside me that day. I pictured myself telling my classmates, "And that's where *I* live." Everyone would *ooooh* and *aaaahhh* with appropriate amazement. And for that brief moment I'd feel important.

I trained my binoculars on the hills to the southeast: *Okay, there's Post Hill in Lyme. A little to the right should be Cobble Hill. Yep, there's the bright metal roof of the state garage, just below our house. And . . . wait. Hey.* Six miles away through the mock-summer haze, barely to the left of Ely Mountain's slope, there it was: the outline of our roof.

Startled, I felt a tingling pinch, like static electricity passed from someone's finger. It wasn't the thrill of self-importance, though; it was geometry. It was as if the imaginary spoke I'd drawn from there to here had become instantly solid—a nylon tether, a belay—supporting me with a reverse view of myself in the valley. Just then, from this strange angle, I could envision me standing on our back deck looking out at this hillside. Even stranger, I could envision me standing on our back deck *being seen* by myself from this hillside. It was a bizarre sensation, as if my reflection in some mirror had, of its own will, just cocked its head and looked back at me.

Now, there was a question: What if your reflection showed some independence? What were you supposed to do then?

Collecting my pack, I climbed out of the clearing, heading north along the ridge. The hike was all wondrous distraction, everything inspired with spring motion. I walked haltingly, like an overwhelmed tourist, taking off my pack, pulling out my field guide at every new wildflower, working to identify it, then making notes and repacking. But as soon as I was packed and moving, I'd see another flower and the whole process would repeat. I wound up carrying the guide in my hand.

The trees were alive with birds I couldn't identify, singing calls I didn't recognize. I was looking down at flowers and up at birds and down at flowers and up at birds, responding as the sight or sound required. It was more than I could handle. So I had to choose: flowers on this hike, birds some other time. I came across yellow wood sorrel, dog and downy violets, gay wings, wild oats, foamflower.

There was little evidence of human presence on the ridge, none since the orange hood, so the structure on a nearby tree brought me up short. Three boards had been nailed into the trunk like ladder rungs; above them was a horizontal platform, painted black. It was a tree stand, a nest for a hunter.

The stand brought to mind the object I'd used as my hub on Ely Mountain. It was a chair, made from two plywood squares—one for the seat, one for the back—set on the edge of a high outcrop wall that ran like a spine along the middle of the summit. Two thin logs of hop hornbeam had been used to form a triangular leg rest. Extending out from the bottom of the seat, the logs came together at a point, ∧, in open air several feet in front of the chair. They were tied together there with yellow nylon rope.

On my last visit to the mountain, I decided to try the seat. It was surprisingly comfortable. I was grateful for the fixed bearing it had given my wanderings, and now for the rest it provided and the view it offered across a wide ravine. Listening to the clacking of two woodpeckers and the chatter of pine warblers behind me, I watched Griffy run excitedly back and forth below me. What had the seat been built for? The same rest and seclusion I was enjoying? Probably not. Or not primarily. Given the remnants of plastic sheeting stapled to the outside of the leg rests, it was most likely a deer blind that could be rigged with a tarp for cover.

For that use, it had a prime location. Any creature making its way through the ravine would have to do what Griffy was doing, cross directly in front of the person in the chair. Sitting there was like being in the first-row balcony for a play, or in the judges' stand on parade day.

A bird's call pulled me out of the seat, away from Ely Mountain, two miles further north on the ridge, several weeks ahead in time, back to this hill, where I was stopped beneath the tree stand. It wasn't just the repetitive two-note song that had drawn my attention; it was the volume. The call was at least twice as loud as any of the others around. And after the third or fourth wave of notes I thought to myself: *oven bird.* What a strange reaction to have had, particularly since I wasn't a birder and had never seen or heard an oven bird before. I'd only come across it when I was teaching Robert Frost's poetry. His poem "Oven Bird" begins:

> There is a singer everyone has heard,
> Loud, a mid-summer and a mid-wood bird,
> Who makes the solid tree trunks sound again.

It wasn't much, but it was all I had to go by. That and the call itself. So as that unseen bird kept belting out its two-note song, I abandoned the wildflowers and tracked the call instead. I followed it downhill past the tree stand, then swerved uphill again when I heard it on the west side of the ridge. Finally, the bird perched long enough in one place for me to lock onto it with my binoculars. Once I had, I felt lucky to have found it at all. Though it was extended in full profile, I could barely see it; its pale yellow-green color camouflaged it almost completely amid the spring leaves.

Almost completely. Everything except the eye. It wasn't anything unusual in itself—just a black circle surrounded by a white ring—but it had a powerful effect on me. Maybe the bird's muted coloring highlighted the eye. Or maybe the black/white contrast magnified it, bugged it out, made it look twice as big as it was. Either way it dominated the entire field of view in my binoculars.

I flipped through my bird guide and, sure enough, it was an oven bird. Yes! Amazing. What a feeling of triumph. Not only had I identified a bird—a rare enough occurrence—but I'd done it without ever having seen or heard the bird before. I'd done it by combining two sources of knowledge—a present sound and a remembered poem. For one of the first times in the past seven years I felt as if my thinking had matured; finally, I was aware of using a hybrid kind of knowledge, a creative way of seeing, just as Ned had with the water drill. Experience of mine outside the landscape had become useful within it.

The voice said, "Before you get too excited, don't forget that you chased that bird all over hell and gone. You're lost now, you know."

"Who cares?" I said. "Did you see what just happened?"

"Oh, I saw."

"You hate that, don't you? You hate it when I do something right."

"Why should I hate it? It's a nice break from the usual. All I'm tellin' you is that instead of runnin' around after poetry birds here on your chirpy Oven Bird Hill, you should be findin' your way home so that you can act like a dad—a real, reliable dad—and explain to your older son why it isn't nice to help his friends shoot each other with arrows."

"C'mon, don't bring that up," I said, my joy suddenly gone. "Not here, not now. Why can't you just let me have a couple of minutes of fun? God*dammit*, you're obnoxious."

"Of course I am. That's my job."

"Well, I'll show you how wrong you are. I'll get us off 'Oven Bird Hill' right now."

I tried to get my bearings, tried to guess where I was in relation to Lake Morey. Looking at the lay of the land, remembering the distance I'd walked this morning, and imagining this ridge's shape as if on a map, I had a feeling that I was pretty close to the lake. So I turned east and headed downhill.

The oven bird followed me. Not the bird itself—it had long since flown away. But its shadow remained, the shadow of that eye, that still, capacious black hole of an eye, swallowing my binoculars, then me and the ridge and the whole sky at once. I began to think the voice had been right to call this place Oven Bird Hill, even though he'd done it sarcastically: the bird's influence on me, I could tell, was strong. But his description of its song as "chirpy" was off the mark—or at least inconsistent with Frost's perspective. Frost ascribed to the oven bird the ability to sing noisily of the world's imperfections, things other birds glossed over with their idealistic chirping—actually, glossed over *because* of their idealistic chirping. Having seen the bird myself, it made sense: if you had an eye like that, you couldn't help but see past everything's bright facade, into its dark interior. And once you did, then you'd need to warn the world about what lay hidden there. How could you not? It would be your responsibility. So you'd sing out as loudly as you could, telling the world about what you'd seen, and from your mouth would come a call of darkness. It was inevitable: a black eye would prompt a dark voice.

Frost understood that well. In a letter to his friend Louis Untermeyer he wrote:

> The conviction closes in on me that I was cast for gloom . . . I am of deep shadow all compact like onion within onion and the savor of me is oil of tears. I have heard laughter by daylight when I thought it was my own because at that moment when it broke I had parted my lips to take food. . . . But I have not laughed. No man can tell you the sound or the way of my laughter.

He also understood that the darkness in his voice was related to the ominousness he sometimes saw within nature. The land's imperfection drew Frost's poetic attention as much as its beauty did. Occasionally, as in the poem "Once by the Pacific," the natural world seemed so overwhelmingly assaultive and menacing that all he could do was call it out, name it:

> The shattered water made a misty din.
> Great waves looked over others coming in,
> And thought of doing something to the shore
> That water never did to land before.
> The clouds were low and hairy in the skies,
> Like locks blown forward in the gleam of eyes.
> You could not tell, and yet it looked as if
> The shore was lucky in being backed by cliff,
> The cliff in being backed by continent;
> It looked as if a night of dark intent
> Was coming, and not only a night, an age.

But more often than not Frost could see past nature's "dark intent," could gather himself and step away and put it in wider perspective. And when he did, he usually treated the land's bruises or gloom or menace not as something to avoid, but as something to wonder at and make sense of. Like his truth-telling oven bird he constantly asked "what to make of a diminished thing," and surprisingly often what he made of it was a connection, a reconciliation. You can see it happen in the poem "My November Guest," which begins with the speaker walking through "these dark days of autumn rain" accompanied by his "Sorrow," but ends with this awareness: "Not yesterday I learned to know / The love of bare November days / Before the coming of the snow."

In fact, reconciliations like this one—between a person and a land-scape—happen so often in Frost's poems that the act of connecting must have been important, maybe even therapeutic, for him personally. It was as if he understood at some level that a person's acceptance of nature in its less-than-ideal conditions made it easier for that person to accept his own darkness and imperfections. Frost sums up the understanding best, I think, in "Hyla Brook," a poem describing a seasonal mountain stream. The description takes place not during spring flood, when it would be easy to wax poetic about the brook, but during midsummer, when "Its bed is left a faded paper sheet / Of dead leaves stuck together by the heat— / A brook to none but who remember long." This is, of course, a set-up: Frost is presenting this natural phenomenon in its "diminished" state as a test. He wants us to ask ourselves, as he undoubtedly asked himself over and over during his life, whether it's possible to embrace so unappealing a thing as this waterless brook, and if so, how. The final line provides the answer. "We love the things we love for what they are," it reads, which I've always understood to mean that truly loving something means loving it for everything it is, even its parched, shadowy, ominous nature.

Loving things for what they are. That was a wisdom Asa Burton could have used in his final years to keep from punishing himself for every-thing he felt he was not. Shit, it was a wisdom I should have been using myself every day.

The one piece of wisdom I was getting from Oven Bird Hill as I started down was a faint directional clue: the indentation of an old logging road. I followed it across the slope to a dirt driveway, where I turned downhill.

The driveway spilled into a paved road that snaked down through a hillside neighborhood, down toward the interstate. At a spot where the road curved left, just above the highway, I stopped to get my bearings again. After a minute I saw that I'd come down about a half-mile short of the lake.

My thighs quivering from the downhill walk, my brain dazed from the heat, I was exhausted. So I opted to head south, back toward Ely, and my car, and home. I turned onto a dirt road, and when I found a good spot, lifted Griffy over the wire fence bordering the interstate, then hopped it myself. After picking through a thin border of trees, I came out into a long-grass gully beside the southbound lanes of the highway, ten feet below their surface.

The sun high, the south wind like a blast furnace in my face, this was not what my fatigue needed. So I stuck close to the trees, walking in the shade as long as I could before I had to rest. Sitting against a tree, I faced the road and ate the fruit I'd brought.

Looking up from the base of the tree, all I could see was a bank of lush grass rippling in the gusts, and above it cars occasionally flashing left to right across my field of vision. Or that's what I perceived. What I *saw* when a car passed was the upper half of one side of it and the profile of the driver from the neck up. It was comically stunted, the scene, and flat, like the candlelit silhouette of shadow puppets. Everyone was moving south, attention focused down the corridor of pavement. No one saw me, or at least no one looked. From their seats I might as well have been invisible.

And that combined oddity—my seeing the drivers sideways and their not seeing me—led me to wonder more about how much of the world we did and didn't see. I knew we couldn't see everything; I knew we filtered what we saw, did our best to accommodate the waves of images washing over us. I could even recognize my own accommodation in the way I'd broken the valley into pieces. But where were we starting from? How broad was our spectrum to begin with?

A moving car probably wasn't the best place to start. Traveling at 65 mph, eyes focused on the highway ahead, a driver's field of view was pretty narrow, perhaps no wider than the windshield itself—a 120-degree angle or so. At so slender an angle the terrain would seem to disappear as it passed the sides of the windshield. So it wasn't really surprising the drivers didn't look at me. Sitting down where I was, I was blocked from their forward view by the contour of the highway and was only visible to them at the moment they passed me. By that time, with their attention drawn to the road ahead, I was out of sight.

But even without a moving car to limit it, our scope wasn't much better. Standing still, using our best peripheral acuity, we were only aware of half the world. What existed behind us might as well not be there at all. And even if we could improve on that, if we could see 360 degrees and absorb every stimulus, we still might miss things. In his book *The Origin of Consciousness in the Breakdown of the Bicameral Mind*, Julian Jaynes argues that our ultimate limiter is consciousness itself. Using a flashlight for his analogy, Jaynes describes just how confined our consciousness is and how difficult the confinement is for us to recognize:

> Consciousness is a much smaller part of our mental life than we are conscious of. . . . It is like asking a flashlight in a dark room to search around for something that does not have any light shining upon it. The flashlight, since there is light in whatever direction it turns, would have to conclude that there is light everywhere. And so consciousness can seem to pervade all mentality when actually it does not. . . .
>
> It is much more probable that the seeming continuity of consciousness is really an illusion. . . . In our flashlight analogy, the flashlight would be conscious of being on only when it is on. Though huge gaps of time occurred, providing things were generally the same, it would seem to the flashlight itself that the light had been continuously on. We are thus conscious less of the time than we think.

If that was true, if consciousness was inherently limited, and enormous pieces of space and time fell outside of my everyday awareness, then a complete understanding of anything—the landscape, myself—seemed impossible. The best I could do would be to reach a complete understanding of everything I could understand, which would only be a part of the whole. What else, then, was out there? What was out in the valley beyond my vision and understanding? What was inside me that I hadn't the capacity to get at?

Those weren't questions I was going to answer sitting below the interstate, so I packed up and started off again, feeling not much better. Trudging the gully for a mile or so, I followed a rise to the base of a tall, domed outcrop. Instead of veering left and walking the narrow grassy strip between the outcrop's face and the road (I had a vision of myself getting skewered by a piece of falling slate), I climbed the dome along a narrow deer path. On top of the outcrop, the ground was surprisingly dry. Lichen crunched beneath my feet. But the slope was in bloom anyway. Bluets, white violets, young cherry and apple trees blossoming. And from out of the crevices in the rock face, clusters of wild columbine. Thirty, forty stems, thirty feet above the highway. I'd had such a hard time finding even one on my walks before today; now suddenly I was graced by this abundance. What a strange turn.

I continued along the deer path, which dipped and rose with the terrain. Finally, I stopped when it looked as if I had a choice of courses: I could go straight downhill to Route 244, turn right, and then trudge up the long incline of the Ely Wind Gap, or I could turn uphill now and climb into the narrow ravine above me, bushwhacking a steeper but more direct path across the ridge.

Fatigue decided: it said shorter was better even if shorter was harder. So I scrambled up the slope. Slate chips slid out beneath me, taking me with them sometimes, leaving me exhausted and thirsty and on all fours as I crawled higher into the ravine.

Halfway up, I found a rock seat beneath a hemlock. Hollowed like a canoe and grown in with trees, the ravine provided welcome shelter and seclusion and a narrow view across the valley toward Post Hill in Lyme. Though I'd never had this exact perspective before, I'd looked across the valley from this ridge so often that I knew the broadside view of the New Hampshire piedmont by heart: the low, uneven cluster of hills in the foreground—Post Hill, Breck Hill, Kenyon Hill—the domes of Smart's Mountain, Mount Cube, and Mount Moosilauke in a row behind them, and in the gaps between their crests, the peaks of the White Mountains in the distance.

But now something looked different. I didn't know why. Maybe it was because of my fatigue or the tunneled perspective from the ravine. Or maybe it was the way I'd seen everything this afternoon as if from the side—the wind gap, the hunter's hood, the oven bird, the passing cars on the interstate, the drivers' heads. Maybe it was just the direction of the wind. But in that moment I thought I had a sense of the valley's own character. I thought I could see its face, its profile in silhouette.

The Face

It wasn't anything hidden. The features were traced right in front of me. I knew them as well as I'd known anything in the last seven years. I'd even gone so far as to chart their direction: from Cobble Hill to the Palisades, the river and the two piedmont ridges lay in a clear north-northeast line—a 20-degree bearing on a map. But sitting in the ravine and seeing that bearing from the side, as if in profile, I sensed its prominence in a way I never had before. It was as if the entire landscape was defined by the river's line.

I didn't know why I hadn't seen it before. I should have noticed it every time the geese migrated up or down the valley. I should have noticed it on those days when the landscape wouldn't yield to the line I tried to fit it with, but left me instead to take what it would give. That had happened in the fall two years earlier. I had walked up Post Hill in Lyme in order to find a different perspective, one looking west into Vermont. But it was impossible; the hillside was so wooded that there was no clear view to the west. Instead, at the summit I came upon a sheer cliff and a view to the south and east, further into New Hampshire.

It was the signature of glaciers, this contour—the northerly slope of the hill graded gently by the advancing ice, the southerly one chipped and blunted by the glacier pushing over the peak—and it determined the direction I was going to look. What could I do? I had no choice but to recognize the hill's own configuration and accept it. So I used a notched outcrop on the peak for a chairback and sat down to rest, tilting my face toward the morning sun.

Now I could see that Post Hill had done to me that morning what the valley had been doing to humans for thousands of years. The legacy of its influence was printed in parallel lines along the valley floor.

When European settlers poked this far north in the mid-eighteenth century, they found what French trappers had come upon a century earlier, and what the Abenaki knew thousands of years before anyone else —that the best road was the river. The Abenaki canoed it in the warm months, walked it during the winter. And their footpaths paralleled the waterways as well—the Connecticut, the Ompompanoosuc, the Ely Wind Gap, as well as the smaller, steeper brooks draining the ridge.

Euro-Americans recognized the worth of the rivers too, but in a different way: for collective commerce. They used flatboats to carry goods on the water and hauled the goods across ice in the winter. For a hundred and fifty years the river served as a log road too, carrying millions of feet of timber to mills downstream.

But the river was capricious. It flooded in the spring and grew so shallow by midsummer that you could hop across it on stepping stones. In the end it wasn't the consistent road that burgeoning American society needed for carriage and wagon traffic, and later for automobiles. So in 1795, according to Frederic Wood in his book *The Turnpikes of New England*, the Vermont legislature resolved to have a public highway built "from the south line of this state to the north line of the town of Newbury," running "as near to the banks of the Connecticut River as may be eligible and convenient." That highway was the ancestor of the present Route 5.

About fifty years later the railroad struck a line through the valley, taking advantage, at least from Thetford to Fairlee, of land between Route 5 and the river. Instantly the railroad replaced the river as the great commercial highway, and all of the villages that had once been river towns now reorganized themselves around the rails.

Finally, a hundred and twenty years after the railroad, the Interstate Highway System arrived, carving with it a hundred-yard-wide swath on the Vermont side, along the foot of the ridge. A friend told me that the valley was filled with smoke for three weeks when the crews began roughing out the interstate's path in 1968. They cut down trees and rolled them into piles and burned them, using tires to set the fires and big fans to spray a steady mist of fuel on them to keep them burning. Two years to rough out the road and a year to build it: by 1971 Thetford was another exit off the interstate, which was becoming this era's river, the path of least resistance, carrying people and goods between the Canadian border and Long Island Sound.

The river, the ridges, the footpaths, Route 5, the train tracks, the interstate: I didn't know whether all of those lines meant that there was a kind of geometry in the valley. I thought probably not. Fritjof Capra writes in his book *The Tao of Physics*, "Geometry is not inherent in nature, but is imposed upon it by the mind." But still, I had the strong sense that the valley had something—an identity, a character, a tendency—something independent of me. And it had to do with all of those lines.

"Sometimes you make me crazy," the voice said. "Why do you always

have to stop and think and dream up frilly ideas about what things mean
and how they go together? Face it, even if things did go together, which
they don't, you wouldn't be able to figure it out. I mean, you're not exactly Einstein. So why don't you just give it up? Stop thinking so much.
See things for what they are. What's your problem anyway?"

"You!" I snapped. "You're my problem!"

"How can I be your problem? I'm the only thing that keeps you goin'."

"You're kidding, right? If I listened to you I'd end up killing myself,
or taking it out on someone else. With you nothing I ever do is right.
I don't know how to be a father, I stink as a husband, can't write a lick.
I'm too dumb to figure things out. Too fat to climb hills. Too many moles
on my face. You told me I had the worst name in the world, but when
I changed it you said I was a traitor for not keeping it. You claim I always get lost in the woods and I'm never prepared, never bring the right
books or clothes or equipment with me into the field. You always tell me
I'm going to get in trouble, even for the smallest things I do. You drive
me fucking nuts!"

"Oh, but it's all bliss bein' around you, right? Listen, if it wasn't for
me, you'd be off in la-la land with all of your little-boy fantasies. You
want the valley to be a wilderness and have a face and keep nice, hidden
wild areas on every hill. And you want rivers to come bubblin' out of
a single pretty fountain at the source. You want kids who don't cry or do
things wrong. You want a wife who knows what you're thinkin' even
though you don't tell her. You want to be the father who can cure anythin', the husband who meets his wife's every need. Shit, all you want to
be is perfect."

"What's wrong with that?" I said. "Better to be like me than the way
you are."

"You *are* the way I am!" he shot back. "Don't pretend you don't know
that."

"I am *not* like you! I hate who you are. I'd get rid of you in a second,
if I could."

"So why don't you?"

"Because I don't know how to do it!"

"That's not why."

"Then why?"

"Because you can't do without me."

"Believe me, I'd be fine without you. More than fine."

"Sure you would. Why don't you tell me what happens when you can't be your dream person or when the world doesn't fit your ideal? What happens when you don't live up to your own expectations?"

"We both know what happens," I said.

"Yes, we do: I take over and do my thing. I get pissed off at Robin or Carry or Jacob, or whatever happens to be in your way at the time, and you feel better. And then when you stop feelin' better and start feelin' guilty about how you got pissed off at them, I come back and work on *you*, and then down, down, down you go."

"Tell me about it," I said. "That's what's been going on this whole afternoon."

"I think you like it," he said.

"I told you, I hate it," I said back to him. "But I think *you* like it. And that's what scares me most about you. You're with me almost all the time, you have been since I was little, but you don't give a shit about me. You never have. You just care about coming in and doing your thing."

"Hey, you're the one who's given me the job."

"Maybe. But maybe not. You know what I'm starting to see, now that I think about it?"

"Go ahead. Bowl me over."

"I'm starting to see why you're here. You're not the savior you've always made yourself out to be. You're not here to protect me from wandering off into my dreamworld. It's totally the other way around, isn't it? I didn't choose you; you came into this world with me. The dreamworld is what's saving me. *That's* what I've created to protect myself from *you*."

"Whoa, that's profound," he said, still dismissive, though with a little hesitation. "Anyway, what's the difference? Still leads to the same psycho guy."

"Nah, there's a big difference," I said, feeling calmer. "It means that you're the psycho, and the rest of me is healthier than I thought."

"Whatever. All I know is this. You need someone to tell you the other side of things, and that job falls to me. Case in point: you've been worrying so much about this stupid 'face' idea that you've totally missed what's been happenin' over your shoulder for the last fifteen minutes. Look behind you."

I swiveled around and noticed for the first time the deepening shadows. The sky slinging over the ridge from the west was the color of a bruise. It had come without warning, or none that I'd been aware of. Here was the dilemma of 180-degree vision again, and here was today's answer to what lay behind it: a storm sky. It was coming now, coming dark and in a hurry, flashes strobing above, low rumbles tagging along behind.

"Holy shit," I said.

"This must be my fault, right?" the voice added.

Thunderstorms had always frightened me. As a child, even in the security of my own bed, I'd felt unsafe. Vulnerable. The thunder sounded like the rage of something much bigger than me, something I was powerless against—an ocean wave that would tower above me, knock me over, sweep me under, hold me down. But now, sitting at the base of this tall hemlock, it wasn't so much the thunder I was afraid of; it was electricity. I was worried about getting lit up like a light bulb.

"About 70% of all fatal lightning accidents involve one person," I'd once read in Martin Uman's book *All About Lightning*. "About 75 to 85% of all lightning deaths and injuries are to men. . . . About 70% of all injuries and fatalities occur in the afternoon. . . . The largest single category of lightning deaths (12 to 15% of all fatalities) is composed of those unfortunate individuals who seek refuge under trees during thunderstorms and have their sheltering tree struck by lightning."

The voice said: "Let's see. One person, a male, sittin' under a tree—and, uh, what time of day is it? Oh, yeah, it's afternoon. Hey, that makes you four for four, puts you comfortably in the high-risk category. Well done! Might as well lash yourself to the top of this tree and tie a golf club in your hand. Then you can wave it around and pretend you're a conductor."

"Watch out or I'll tie *you* up there," I said.

I tried to act unconcerned, but that was a sham. Inside I was thinking desperate things, like freeing myself from all of my metal: my belt, my wedding ring, my necklace and pendants. How about the lace loops on my hiking boots?

"Don't forget the zipper on your shorts!" the voice added.

No, this was ridiculous. I had too much pride to start stripping down, even if no one was going to see me. If my time was up, I was going to go with my clothes on.

In minutes the wind was gusting. Around me the trees groaned and clacked against each other. Lightning blinked like camera flashes. Thunder cracked above me, sometimes right overhead, sometimes farther away. Then rain. Clouds of rain, clogging the valley like fog. Wind, lightning, thunder, rain, all of it moving fast: it was a kind of storm we didn't often see around here. Though I slid away from the trunk a bit, I sat out the storm nervously under the hemlock's protection. Griffy did too, though he kept his distance from me and my metal.

After twenty minutes or so, the air quieted. And when I could see that the bolts of lightning had moved beyond the New Hampshire piedmont, I pulled myself together and hiked up the slope. It was the steepest climb of the day, through mixed woods, much farther to the crest than I'd expected. As the terrain leveled out, I noticed the outline of the wind gap ahead, and through it a brightening sky to the west. To the right of the gap, silhouetted against the white horizon, rose the rounded end of Oven Bird Hill, the place I'd started, still a mile away.

I bushwhacked through the woods until I crossed a dirt road, a driveway that I followed as it looped sharply downhill past a room-sized carpet of bluets, more bluets in one place than I'd ever seen. Turning right at the end of the driveway, I walked uphill past a couple of houses and into a smell that was lush and thick as a curtain, sharply sweet. I stood still for a minute, to catch the quiet there on the road, broken only by the sound of dripping leaves. Then I breathed the scent in again and again. It was familiar but elusive; there was apple blossom in it, and pine, both swollen with dampness. There was more to it too, but I couldn't get a bead on the rest. It was too subtle.

As the road leveled out, then descended, I caught sight of Craig's house up ahead. A little further on I found him in his garden, right where he'd been when I started uphill earlier that afternoon. We waved. His shirt off, he looked calm and dry, as if there'd been no storm, no excitement.

On the way home, though, I could see just how much action there had been. Halfway down 244, I passed a large tree in jumbled, chain-sawed pieces on the shoulder: it must have fallen across the road during the storm. In a yard next to Route 5, a six-foot-high well house lay on its side, not far from a downed tree limb. A hundred yards south, an antique shop's hanging sign had been blown half down and now dangled like a loose arm from one of its two hooks. Everywhere there were tree pieces:

the village lawns in North Thetford were littered with debris. When I stopped at the village store to buy a drink, the woman at the register told me how the storm had come slamming through town from the north, blowing everything down.

That was curious. At his place Craig had looked as if the disturbance had simply been a gentle shower. Tucked into the shelter of my leeward ravine, I'd felt a storm with sound and speed. Down here something more powerful than either of those had lumbered through, swatting the village with its paw. But it had all been the same storm, hadn't it?

Standing in front of the store, I drank a bottle of lemonade and looked north along Route 5. Everything seemed to be in pieces: the trees were broken and scattered, my geometry of the valley was in shambles, my conception of myself had been tossed around too. So I didn't see why a storm had to be any different. Maybe a storm or a landscape wasn't a single pattern at all, but a combination of patterns melded to form something that looked whole, indivisible.

That could make sense. I knew *I* felt like that sometimes.

If the world really worked that way—if unities were really complexities disguised—the challenge would be to learn how to look in directions you normally couldn't, to see through the apparent singleness into the disparate features below. And if you could do that, even for a minute, you might really be able to discern the truth about things, or at least the discrete elements of that truth. I might finally comprehend the essential nature of the valley or the unchangeable core of my self.

Back at home, when I walked through the door, I thought I heard a hush come over the house. It had started happening lately, that hush. I didn't know for sure what it meant, but I had a pretty good idea: it had to do with my moods.

In *Essential Papers on Depression*, James C. Coyne writes, "A brief interaction with a depressed person can have a marked impact on one's own mood. . . . Depressed persons often unwittingly succeed in making everyone in their environment feel guilty and responsible." The effect on family members, most researchers agree, is even deeper. Coyne continues, "The interpersonal difficulties of depressed persons are less pronounced when they are interacting with strangers than with intimates."

I didn't know that, of course. I hadn't read anything about depression then and had only begun to crack the denial that I'd been using for protection my whole life. But I had my own picture of what was happening during that hush: Robin, Carry, and Jacob were holding their collective breath, preparing themselves for whatever mood I'd be in. Would I be happy? Sullen? Withdrawn? Grumpy? Angry? They had to wait until they saw me or heard my voice to know how to react.

It was devastating to think it had come to that: my family was scared of me.

Robin had tried telling me about it—an act of immense courage, I now realize, since she often had to endure my defensive anger in response. Sometimes, though, she'd wait until I seemed more able to listen.

"It's hard to walk on eggshells around you so much," she'd said. "It's exhausting. I can't do it much longer."

"But I haven't raised my voice as much these days, have I?" I asked.

"It's not just your voice, Ter. Even your silence can be scary. Scarier than your voice. It still has that aura, like you're mad at everything, but you just do it without words. And I feel like I'm the prime target. And I don't deserve it."

"But it's not about you. It's about me," I said. "I wish I could explain it."

"Why don't you try? Tell me what's going on inside you."

"You don't want to know."

"What do you mean?"

"You don't want to know what's going on inside me."

"Why not?"

"Because I think it'd scare the shit out of you."

She stood there, silent, uncertain, looking at me.

"It doesn't mean I don't love you," I said. "You know that, don't you?"

"Yeah, I do. But that just makes it harder, you know? I don't understand how you can say you love me and still treat me the way you do sometimes. I'm not perfect, but I know I don't deserve the disapproval I get from you."

I looked down. "You're right," I said. "You don't."

"I don't think you mean to do it, Ter. Or not all the time. And that's why it's probably some kind of depression. Why won't you go see somebody about it? Just see what they think?"

"It's hard," I said, my voice low. "It's hard to think I need help. I hate to ask for it. I *hate* it. I just can't figure out why I can't handle this on my own."

"No one's perfect," she said. "No one can handle everything. You want to be so self-sufficient, but everyone needs help sometimes. And you don't want to keep feeling like this, do you? Please tell me you don't, because I can't do this much longer. And we won't survive like this. We may love each other, but we're both miserable. You're angry and I'm worn down."

"Well . . . I need to think about it, okay?" I said, putting my arms around her, and for a while we'd stood quiet, our bodies tight together, feeling each other breathe.

That was the last we'd said of it, and I had tried not to think about it again. Until tonight, when I walked in the door and thought I heard a hush.

So as I moved through the mud room and into the hallway and prepared to turn into the dining room, I thought I'd find wary, expectant faces turned toward me.

But instead, I stepped into a Norman Rockwell painting. Robin, Carry, and Jacob were all sitting around the table, eating dinner, looking flushed, happy. When he saw me, Jacob, in his high chair, kicked his legs and banged his hands down excitedly. Then he reached for me.

"Hi, Dad," Carry said and smiled.

Robin raised her eyebrows playfully. "Daddy, you look like a drowned rat."

"Hey, everybody," I said. "I got caught in the rain."

"It was some storm, wasn't it?" she said.

I let my pack drop heavily to the floor. "It was indeed," I said. Walking around the table, I gave Jacob a kiss on the forehead and Robin one on the lips. Then I walked back to Carry and put my hand on his shoulder. There was unfinished business.

"So, Care, what happened this morning?"

"What?" he said guiltlessly, his mouth full. I could see he wasn't pretending; he really didn't know what I was talking about.

I looked over at Robin; she seemed oblivious too. In fact, the whole mood in the house was lighter than it had been at noon. The bow-and-arrow mishap, like the storm, seemed to have passed over and moved on.

But it hadn't yet for me; I'd been living with it, and the voice's antagonism, all afternoon. I had to get it off my chest. I asked, "Why did you take your bow and arrows out this morning when you knew you weren't supposed to?"

"Oh." He turned his face back down toward his plate.

"Did you just forget the rules, or what?"

"I don't know, Dad," he said and glanced up at Robin.

"You know," she said, trying to intervene, "Carry and I talked about it after Laura, Wendy, and Will left, and I think he understands now that . . ."

"But I want to know what was going on inside his head," I said to her, my tone sharp. Then I turned back to Carry. "What on earth were you thinking?"

"Dad, I'm sorry."

"Ter. . . ," Robin said.

"I . . . just . . . want . . . to know!"

It was quiet for a minute—I standing next to Carry, staring down at him, he looking back and forth between his plate and Robin.

He said, "I think . . ." Then there was a long pause; Jacob made a spitting noise into the quiet. He said again, "I think . . ." and then in an overblown pantomime tapped his temple dramatically with his finger, while he rolled his eyes in circles.

"What?" I asked, surprised. "What's that mean?"

"It means . . . ," he said, doing it again, "I think my head was having a bad day."

Before I could manage any kind of response, the voice erupted in laughter, "Oh, man! He got you there! Zinged you! Never saw it comin'! Wooooo-hooooo!"

"Shut up!" I said back to him. "I'm trying to be serious."

"It's too late. I can hear the seriousness drainin' right outa you!"

It was true. Before I could decide whether I was mad or insulted, I found myself squeezing my lips together to keep from laughing. The shadow of blame and responsibility I'd been carrying around all day was dissipating. It was as if Carry's words had drawn open the curtains on the unlit room inside of me.

"Your head was having a bad day?" I asked, half seriously.

Maybe I wouldn't have been as amused if I'd thought he was making an excuse. But I could tell he wasn't. He was trying to be sincere now, trying to explain his mistake as best he could. And he was trying to make me laugh.

Wonder who he got that from?

The voice said, "Hey, loser, you were right for once. That's a great kid!"

I rolled my eyes at Carry. "Ohhhhhh, your head was having a bad daaaaay."

"It's true!" he said.

"Mmmmm hmmmm," I answered, mockingly unconvinced. But as I walked behind him toward the kitchen, I tousled his hair to let him know that I really did understand.

I mean, a bad head day: who hadn't ever had one of those?

Chapter 6

To the Well and Back

I

On a June morning, three months before Jacob was born, two men—one carrying a tightly rolled map under his arm—stopped by our house. They were looking for water.

East Thetford's old municipal well, a shallow spring-fed one down near the village, had been contaminated, they said: road salt from Route 113, coliform from who knows where. The water in the village wasn't safe to drink anymore. So here they were, a mile uphill, scouting out a new source.

We talked about our well and the other ones in the area for a while. Then, the one man unrolled his map on the hood of my car, spreading his hands across it the way you'd smooth a creased bed sheet. It was a white photocopy with grey boundaries and short dark lines across the middle. Petting the slantlines with his finger, he finally pointed to the darkest one.

"This is it," he said.

A fissure in the rock, a great ledge sluice, a rich vein of water running somewhere under Cobble Hill. The solution to East Thetford's problems. If the map was right, he said, the crack wasn't far from the southern edge of our property, where the land rose steeply. Maybe right below us.

That prospect was enticing, I admit. Ever since we'd moved into the house, I'd been uncomfortable with our water, or with the *idea* of it. Our well was a shallow one, sited in the woods behind the sugarhouse, the lowest corner of the property. It was about fifteen feet deep—five cylindrical concrete tiles stacked one on top of the other. The top eighteen inches poked out above ground and were covered by a cement lid that had an iron eyebolt sticking out of its center. Draped over this exposed top was a tattered sheet of black polyethylene; its edges were pinned to the ground by rocks.

We were told, when we first looked at the house, that the well had provided water throughout the driest summers, even when other people's drilled wells had failed. That was unusual for a shallow well, and a good sign about the supply.

But there were questions. The water was iron rich, which wasn't surprising for this area, but it lead me to wonder about other things getting into the water, a possibility confirmed by a water consultant who knew our well site. "Hey," he said, "it's at the lowest point on the property. *Everything* leaches down there. Bacteria and all."

"You're probably poisoned already," the voice said. "Of course, you won't find out until fungus starts growin' on your face, but no one'll notice that anyway."

"That's pleasant," I said, sounding unfazed. But we both knew I was paranoid that something like that might happen.

This instinctive doubt about the quality of our water was an interesting change for me; as I'd discovered over the last several years, my early response to a situation was often to create an idealized expectation that reality couldn't possibly fulfill. In this case it would have been something about a pure, plentiful water supply beneath our back yard. Dreamy as that sounded, there actually was some historical support for the idea. For instance, in 1794 Samuel Williams wrote in his book *A Natural and Civil History of Vermont*: "In the plains, hills, and mountains in this part of the continent, there is scarcely a place in which water may not be found at the depth of thirty or forty feet from the surface of the earth. . . . the earth at that depth is well saturrated with water; nor does it fail nor is the temperature of earth at that depth much affected, in the hottest, or in the dryest season that we ever have."

But the difference this time was that I'd seen the ideal betrayed. It had happened five years earlier, while I was writing an article about hazardous waste in Vermont. I talked to Clara and Tom Bolton, who lived in the southern part of the state and had drilled a deep well in 1962.

"We had it tested and analyzed by the state. It couldn't have been better or purer water," Clara said.

Seven years later their pure water had become putrid. The smell was so bad that they stopped using it and instead carried home what they needed from a community well, until the smell turned up there too.

New tests done on the Boltons' water showed dangerous levels of volatile organic compounds and of metals such as mercury, cobalt, cadmium, beryllium and selenium. The chemicals ate away at the enamel in their tub and the nickel plating on their faucets. The substance oozed from the taps in a jelly.

It turned out that various chemicals, along with the town's other waste, had been poured into the ground across the street from the Boltons years before, when that property was the town dump; over time the chemicals had percolated underground into the water supply. There had been no way for the Boltons to know.

Clara said she had had cancer, Tom still had circulatory problems, and her granddaughter had been born with birth defects. She blamed the water. To find water they trusted, she and Tom now traveled seventy-five miles every week to a clear spring, where they filled as many bottles as they could fit in the back of their car. More than the rest of us, the Boltons had come to know the weight of water.

I talked to other people whose experiences were similar to the Boltons'. I saw the elaborate filtration systems they'd installed in their houses and the amount of time they devoted each day to monitoring the quality of their household's water. And I tried to imagine how different life would be if that suddenly happened to Robin and me.

Those people's experiences reinforced the grand mystery behind this ideal of water in the ground: its invisibility. If the intricacies of groundwater movement through rock—the faults and folds, porosity and permeability, pressure and gradients and gravity—happened out of sight, how could you tell for sure whether there was a river the size of the Connecticut under you, or just parched gneiss?

That's what the man with the map was doing at my house—trying to answer those questions about East Thetford's groundwater supply. He had a methodology for locating it too; it involved examining the terrain, talking to people and analyzing different kinds of data. Once he'd done all that he felt he could make a very educated guess about where to put the new well. From talking to him that morning, I could tell that what he had in his mind's eye was the kind of bounty Williams had described two hundred years earlier—a stream, a river, a water-rich crevasse hundreds of feet underground.

That was fine. But even if he found it, I kept wondering, how could he keep the people down in East Thetford from the Boltons' fate, how could he tell them for sure what was coming from across the street, or from some direction they hadn't been looking? How could any of us protect our children or ourselves from deadly water?

That was enough of the questions: I had to see inside our well myself. Then I'd know for sure how safe we were.

One day I marched over to it and kicked away the stones. Pulled back the plastic cover. Squatted down and pushed against the rim of the concrete lid. Nothing. Leaned over, wrapped both hands around the iron eyebolt in the center of the lid and pulled hard. Still nothing. Threw myself against the rim. Nope. Tried each method again in turn, but with no more success than before. Not an inch.

"Need some help?" the voice asked.

"It's a lot heavier than I thought," I said, breathing hard.

"That's surprising. It's only a four-inch-thick slab of CEMENT!"

"So how am I going to move it?"

"You're not, unless you get some help."

"And how stupid would that sound?"

"Well, let's see," he said and then made a little-boy voice, "*Mister, could you help me move the lid off of my well? Pretty please? I wanna look inside and see the wa-wa.*'"

"I guess *that* ain't happening."

So instead of sounding boyish and needy, I surrendered to the lid— pulled the plastic back on and pinned the corners down with the stones. Then I decided I wouldn't worry about the well. I was through with it. I didn't really need to see inside anyway. What did I think I'd find? Cider-colored water? A mini Love Canal? A swirl of jellied microorganisms? Nah. Forget it.

Instead, I vowed to eliminate the whole issue and drill our own deep well sometime soon. *That* would take care of it. I was sure it would. Even the water consultant, who had doubted the purity of our shallow well, had said it would. In the meantime, if we couldn't see the water in our well, then, by god, we'd use as little of it as possible: we'd buy gallon jugs of water for cooking and drinking. And I'd put that damned thing out of my mind.

Hardly.

The truth was, I couldn't escape the uncertainty. I kept wondering what it looked like inside, what our water looked like. Sometimes when I was over on that side of the property I'd sidle up to the sugarhouse and peek around the corner at the well, dark and stunted and wrinkled. It looked like a gangrenous wart in the woods.

2

On a summer afternoon ten years before the man with the map knocked on our door, while Rob and I were still living in Chicago but visiting out east, a friend named Bushy took us to a waterfall/swimming hole near the base of Mt. Washington in New Hampshire. It was hot; he said he knew a good place for a swim.

We stopped at the cliff's edge opposite the top of the falls and looked down thirty feet to the pool. It had rained hard earlier in the week, and the waterfall now broke full and mean over the edge. Below, the pool churned with white foam.

"I've never seen it this strong!" Bushy said, his eyes wide and excited. There was more than swimming on his mind.

Pulling me to the edge, he pointed. "The key to the jump is to hit the right spot. If you jump out too far you'll hit those rocks. If you don't jump far enough, you'll either hit the near ones or get swept under them by the current. You can't see it but there's kind of a shelf there and you can get caught underneath." He bent me over the edge to show me where he meant. "So you gotta hit the middle."

Wait. This was supposed to be a swimming trip. No one had said anything about jumping.

Two or three other people were sitting on the cliff admiring the rush of water. A rumor circulated through them about someone who'd been hurt recently making the jump.

"I wouldn't do it," one of the observers said.

That was testimony enough for Robin. She may have had the breath of adventure in her, but she wasn't stupid. She was not going to jump and tried to talk me out of it too.

"Forget it, cliff diver. It's too late," the voice said. "You had your chance to say no, right away, when he was showin' you the jump. You didn't, so now you're committed. You back out now and you'll look like the biggest wimp in the world."

He was right. The issue was no longer rational; it had been taken over by my swollen male ego. Simple, overwhelming pride. If Bushy jumped, I had to too. No question.

And he did, a perfect feet-first jump, knife clean into the center of the pool. He bobbed up and swam easily to the far edge.

Let me be clear about this: despite the prodding of the voice and my own machismo, despite the calculations I'd made in my head, despite the confidence that I knew in strict geometric terms where the center of the pool was, despite the fact that I'd seen Bushy hit that spot like an arrow in a bull's eye, I did not want to jump. What worried me most was not being able to see what I was jumping into: the pool surface was clouded by bubbles. I kept imagining myself held down by what John McPhee called "the great, almost imponderable weight of water, enough to crush a thousand people." I guess what I imagined was my own death.

I wish I could say that what I finally did was inspired by a profound, otherworldly connection to the place, or some epiphany, some moment of transcendent knowledge or insight, a clear intuitive vision of my own survival that willed my body into the center of the pool. But there was nothing that dramatic. Just pride.

And so I jumped.

I remembered the leap clearly, even ten years later as I worried about our well water, because there was that instant of electric, goosebumping fear, and then surrender. It was a strange feeling, that first airborne moment: having tried to see and know everything myself, control my own trajectory, I finally had to behave as if I didn't care. Heavy as a stone and dropping fast, all of my best intentions were reduced to an act of faith.

3

Eight months after the two men visited our house, five months after Jacob was born, I followed through on a deep well. It was February when Bill Bailey brought his percussion drill over and set it up. Since the ground was completely frozen, he couldn't easily dig into it to lay supply pipes, but he could still drill. As long as we didn't need the water right away, he said he'd come back and lay the pipes in the spring, after the ground had thawed. That wasn't the best situation, to my mind: it would have been nice to have the water right then. But we'd survive.

I'd never known there was such a thing as a percussion drill. I'd assumed that well drillers just used giant versions of the kind of hand-held rotary drill I used around home, the kind whose raised, beveled edges cut the hole as the drill bit spun. Some drillers did use that kind, I found out, but others, like Bill, used one that was more like a hammer—a long, dense, cylindrical bar suspended by a cable. When the rig was running, the cable would move back and forth—retracting a little, then extending, retracting and extending, over and over—and when it did the bar would rise and fall, rise and fall, rise and fall, banging, banging, crushing the rock beneath it, working its way further and further down. For days, with the rig standing in the front yard, our house and lives rang with that rhythmic hammering.

Before siting the well, I had taken a cue from the man with the map and researched as much as possible. I located our closest neighbor's septic system and tried to imagine whether their effluent would percolate downhill toward our well. I asked other neighbors about the depth of their drilled wells and the quality of their water. On our own property Robin and I had three criteria: we wanted the well close to the house (laying the supply pipes would be less expensive that way), at least a hundred feet from our own leach field (as required by town health regulations), and as out-of-sight as possible. Finally, we made sure we weren't interfering with the oil tank buried next to the house.

Then it was a matter of picking the exact spot. After all of the research I'd done, I'm embarrassed to say that I chose it with no more care than an impatient child. Not once in that final stage did I seriously consider the landscape itself—the rock or soil type, the topography. Maybe because it seemed so unknowable. In the end, choosing the site was like playing pin-the-tail-on-the-donkey: I closed my eyes and pointed.

I know now how naïve that was, how easily that hole could have come up dry, how much money it would have cost us to drill again somewhere else. Luckily it didn't work out that way. After a week of drilling, Bill found a good supply of water. Of course, without supply pipes we couldn't use it, but the water was there anyway, forty feet from the house.

That fact became more tantalizing a week later, when we returned home from a weekend away to find no water at all. The faucets—all of them—dribbled, then hissed, then went silent and dry. Once flushed, the toilet tanks didn't refill.

Having never been without water before, and not knowing how long we'd be without it now, and imagining Carry and Jacob gasping with thirst, going bathless for months, I panicked, and in the morning left hysterical messages with two local plumbers. But it had been a frigid, snowless weekend, one of many that winter; the frost had crept a little deeper underground, a little further into people's houses, and the plumbers had plenty of other pipes to thaw.

Frozen pipes was one possibility—the best one, from a homeowner's point of view, since frozen pipes could be thawed without damage. Burst pipes was another possibility and not so good, because burst pipes leaked water—usually lots of water—and caused other damage as well. But our ceilings and walls and floors were dry and intact, so we guessed that the pipes still were too. The worst possibility was a dry well; a dry well meant no water until spring, when the ground thawed.

After trying the plumbers, I called Bill Bailey and explained the problem as best I could. He listened, then said, "I bet somewhere between the house and the well them pipes is froze." But he thought we ought to check inside the well first, just to be safe.

When he got to our house, he pulled an iron bar from the back of his truck. With the leverage of that bar stuck through the center eyebolt, we slid the well lid halfway off. I was embarrassed at how easy it was to do.

It was a clear morning, cold as glass, with the sun still low behind the pines. The woods around us were in shadow, and it was even darker and colder inside the well. When I bent over the crescent opening and looked in, all I could see was the light of the reflected sky and my own silhouette. So, there was water.

How much, I couldn't tell. The water was all surface and reflection, all shadow and light. It gave no hint of depth, and we hadn't brought anything to measure its level with. The only sign I could make out was a dark ring around the well three feet above the water line. If that meant anything, it was that the well wasn't at its high water mark. But I didn't know if that was a normal February depth or not.

But the well was not dry. That meant we probably had frozen pipes. Back inside our basement Bill showed me a way to get whatever water was in the pipes moving. He turned our water pump on for ten to fifteen seconds, then shut it off for that long. On, then off. On, then off. Five or ten times before giving the pump a rest. He said he was trying to create a surge of water to break through the ice dam in the pipes. A water percussion drill. I tried it a number of times, imagining the water banging into the ice when I turned the pump on, then withdrawing when I shut it off, little waves crashing into a frozen breakwall. If the water could pierce even a pinhole in the ice, he said, we'd have our supply back.

I spent the next four days trying the pump-surge method, and in the meantime solved our water needs the only way I knew how: I carried water in from the well.

With a two-and-a-half-gallon plastic bucket and a spool of clothesline, I hauled water out of the well and carried five gallons at a time to the house. Actually to the washing machine, the cleanest, soundest, and most accessible reservoir inside. I filled it full every evening after work.

Whatever water we used we scooped from there; any hot water we wanted we heated on the stove. We poured a pitcher of heated water over our heads and called it a shower. We dipped a wash cloth into a pot of water and said we'd bathed. We flushed the toilets only when our sense of hygiene overcame our need to conserve. We didn't wash any clothes. We used as few dishes as possible. Even so we used as much as thirty gallons a day.

In *A Sand County Almanac*, Aldo Leopold wrote, "If one has cut, split, hauled and piled his own good oak, and let his mind work all the while, he will remember much about where the heat comes from, and with a wealth of detail denied to those who spend the week end in town astride a radiator."

The same can be said for water: when you carry it in by the bucket, I assure you that you remember how it got there. A gallon of water weighs about eight pounds, which means that we were using—and I was carrying—250 pounds of water each day. Suddenly, I found myself thinking about water in terms of its weight, couldn't believe how casually we'd used it before. A five-minute shower: 250 pounds. Toilet flush: 40 pounds. Dishwasher load: 100 pounds. Washing machine, full load: 250 pounds. On average, each of us uses up to 800 pounds of water each day, and I suspect we do because we don't have to carry it.

Toward evening on the fourth day, the fourth day of lugging water, I was mindlessly working the pump-surge method when I heard water trickle in the pipes. Or thought I did. Was I hallucinating? Nope. It was a trickle. Definitely a trickle—the sound of little glass beads sprinkled on a dinner plate. Slowly the pump stopped its strained whining. The needle on its pressure gauge—the one frozen at zero for days—moved. We had water again!

In four days, I had come to know water more intimately than I ever imagined I would—as a necessity and a luxury, as a function of gravity, as manual labor. In the months afterwards, I did the best I could to conserve, buying low-flow shower and faucet heads, and dams for the toilet tanks. I promised I wouldn't take it for granted again. That night, though, I turned the shower on and splashed around in it like a kid under the spray of a summertime sprinkler.

4

As is true for other creatures, humans' interactions with the world are defined by the limits of our senses. We haven't the pinpoint eyesight of eagles and owls, nor the echolocation ability of bats; as Diane Ackerman points out in *A Natural History of the Senses*, we don't communicate with the ultrasonic frequencies praying mantises do, nor the infrasonic frequencies of elephants. Still, we've learned ways to improve our perceptual ability: "We're skilled extenders of our senses," she writes.

We love to extend them with tools. We use binoculars and telescopes to discern what our eyes miss at a distance, night vision goggles to see what we can't in the dark. We use long-range side-scan sonar to map the topography of the sea floor, ultrasonic waves to see a fetus in the womb. We position antennas to listen for patterned messages from deep space.

But there are other ways to extend our senses, which is something I learned from geologist Ken Bannister. I sought him out because after our deep well had been drilled, after we'd found a water supply that was better than most of our neighbors', and clean too, I felt even more awed by the mysteriousness of groundwater. I was curious to learn how a professional went about locating it, how he could know, even trust, the existence of something when he couldn't see it. Ken, I'd heard, could find water. In fact, he was at the time, and may still be, the most effective water finder in Vermont: in more than twenty cases he had sited a town's most productive well. In Fairfield, where they'd been looking for a community water supply for more than two decades, and where the average yield of a well was less than two gallons per minute, Ken sited a well that yielded one hundred gallons per minute.

Rolling out his master map of the entire town of Fairfield, a twenty-five-foot aerial map on which he plotted all of his information, he explained the process of finding water. In general, it was a methodical, scientific process, involving layers of information—geological and topographical, historical and hydrological—sifted through progressively smaller sieves of criteria until the only things remaining were potential well sites.

What Ken did first was look at driller's logs—the accounts of the wells drilled in town—to determine where the most and least productive wells were. He'd plot them on the master map. If he was lucky the logs might mention something about site deposits, hinting where old underground valleys may have been. If so, he marked them on the map too.

Next he did a fracture trace analysis. Using a stereoscope, which makes the flat landscape on a map stand out in relief, he looked at infrared aerial photos and tried to discern the structural features of the bedrock. He was interested in indications of fractured bedrock, since fractures could be waterbearing.

"What you look for are tonal changes in vegetation," he said. "A lot of times when you have a fracture trace that runs across a field, the vegetation or the soils above the fracture will be slightly different than the surrounding areas."

After that, Ken consulted a bedrock map, looking for formations of coarse-grained, brittle rock that might be good aquifers. Those formations, along with the fracture trace results, got plotted on the master map too.

You can imagine the clutter on the map—dots marking wells, slant-lines for potential bedrock fractures, large wobbling ovals around various formations, the marks made in several colors, each with its own significance. But it was only because of the clutter that Ken could narrow down his site choices and head out to do the fieldwork.

Once on site he used two instruments. The first was a magnetometer, a device that measured the total intensity of the earth's magnetic field, which varied according to rock formation. Voids in rock provided different readings than solid rock, and voids were what he was interested in. Using a magnetometer, he said, was like drawing a map of the underground topography.

The second was a VLF (very low frequency) receiver, a hand-held device that responded to low-frequency waves that a number of organizations, including the U.S. Navy, sent into the earth. Once calibrated, the receiver made a constant whine; when it went quiet, it suggested an unusual feature underground.

Traversing the entire area twice, first with one device then the other, Ken took readings every fifty feet. Back at the office he plotted those readings on separate maps.

With the data collected, Ken was ready to pick out specific well sites. And it was what he did then to find the exact drilling spot that really surprised me, taught me the most about seeing. After analyzing driller's logs, trace analyses, site deposits, bedrock formations, and readings from the magnetometer and VLF receiver, just when I'd have expected him to crunch numbers the hardest and apply his sharpest analytical skills to the data, he left his office. He headed back out to the field. And he dowsed.

Dowsing started for him years earlier, at Sugarbush Resort, when a dowser was called in to confirm Ken's finding and did: he dowsed the exact spot Ken had flagged. It wasn't the first time, either. On a number of previous jobs a similar thing had happened. But this time Ken was curious.

"How'd you do that?" he'd asked the dowser.

Handing over his rod, the dowser said, "Just ask it where the water is."

And when Ken did, he felt the rod go down right on the spot.

Since then he'd become a devout dowser, even converting some of his colleagues. There was something very real to it, he believed, something psychic or parapsychic; he never pinpointed a well site anymore without dowsing for it first. And the way he talked about it, I got the impression that dowsing was the method he had the most faith in, the one he'd keep over all the others if he had to choose.

"Long enough in the desert a man like other animals can learn to smell water," Edward Abbey wrote. "Can learn, at least, the smell of things associated with water—the unique and heartening odor of the cotton-wood tree, for example, which in the canyonlands is the tree of life."

In other words, we can extend our vision without external artifice. With a little internal retooling we can overlap our senses and go beyond the sensory and intellectual to the intuitive, where our sharpest percep-tions, our most reliable knowledge, may come in inexplicable ways, where we may learn new things without analyzing them at all.

Even with his degree in geology and his elaborate tools—the fracture trace analyses and magnetometer and VLF receiver—Ken wouldn't stake a claim to water until he could see it without his eyes. Holding a rod in his hands, he'd ask it a question, start walking, and wait for a tug.

5

Almost eighteen months after the man with the map visited our house, excavation crews started burying the last section of pipe for East Thet-ford's new well. By that time, Carry, who was five and a half, had just begun kindergarten; Jacob was a little more than a year old and was toddling around confidently; Robin and I had celebrated our ninth an-niversary the month before. Life rolled on as the backhoes and dump trucks rumbled up and down Asa Burton Road over the season's first snowfall.

Whether the man found the fracture he was looking for that June day I don't know. But given that he tapped a dream supply in the woods a quarter mile from our house, and given that he was probably Ken Ban-nister (he definitely worked on the project, but neither of us could remember for sure whether he'd been the one to stop by our house), the odds were pretty good. The well was three hundred feet deep and yielded seventy gallons a minute, more than enough to satisfy the thirty places in the village that would draw from it.

Our well turned out to be a gusher too: one hundred ninety feet deep and more than fifty gallons a minute. There was so much water that it flowed out of the top of the well and back into the ground again, a supply big enough to satisfy the whole village of East Thetford itself, which led me to believe that the man with the map was right all along—there was a crack in the rock, and it bore lots of water, and it ran under our yard and it must have extended diagonally uphill beneath Asa Burton Road and the peeper swamp and the site where they'd drilled the municipal well too.

If so, it meant we'd tapped into that same crack. But if we did, it was by the rawest of luck—though I liked to imagine now and then that I had some Ken Bannister in me, that I'd selected the spot with an intuitive hunch I didn't know I had.

But the truth was, that was a dream. Ken's ability confirmed to me my own limitations—that, like most people, my senses usually stuck to the things they did best: my ears did the hearing, eyes the seeing, nose the smelling, and so on. And when arranged together, that was it: the circumference of my senses determined what I could know.

I might have been able to live with that if what was happening inside that circumference, the rhythm of my learning, didn't feel so dull-witted, so frustratingly inconsistent and vacillating—a two-steps-ahead, two-back pattern, and not the smooth, forward progression I thought it should be. One moment I'd have a streak of enlightenment but then follow with a startling ignorance—as when I sensed the danger of that foaming waterfall pool but leapt in for no better reason than pride, or when I researched our well site diligently but then chose its exact location with careless naïveté, or when I marveled at the glinting surface of well water on a bright winter morning but then made my own reflection more important than the water itself. "So it is with all men," Howard Elman thinks in Ernest Hebert's novel *The Dogs of March*. "Wise here, stupid there."

Maybe so, but it was hard for me not to get depressed about those inconsistencies. They seemed so much like failures. Failures I didn't see anyone else making. Just me alone.

And in the pall of the bottomless moods that followed that lonely sense of failure, with the voice and me shouting at each other, I'd see an image of myself shuffling pitifully in place, like an old man on toothpick legs, not ever getting anywhere, not ever knowing anything better than I had before, not ever understanding the invisible features of the valley or the unrecognized contours of my self, not ever really connecting with someone else.

And perhaps it was that persistent vision of myself—decrepit, dead on my feet—that finally scared me. Or maybe it was the way the voice kept speaking from my mouth, whether I liked it or not, berating Robin during our arguments, saying things I couldn't imagine saying myself. Or the buzz of exultation he felt after he'd gotten her to surrender and cry, making her feel guilty and self-conscious. Maybe it was all of those things together, the way they left me feeling incapacitated, deadened— like a switch, broken and useless, rattling around inside of a running motor it once controlled—that scared me enough to finally accept what I wouldn't before: that I could not restrain the voice, or solve my moods, myself.

"I think you may be right," I said to Robin one day. "About the depression, I mean. And needing help."

"Oh, my god," the voice said disdainfully. "You do *not* need help. What you need is to gut this out on your own."

"What can I do?" she asked, in her face a small brightness I hadn't noticed for a long time.

"I don't know. How do I go about finding someone to see? I don't even know where to start."

The voice said, "Someone to see? You're gonna find some stranger to talk to and tell your deep, dark secrets to? And you're gonna pay them a million bucks an hour for that? Like *that's* gonna work. You can't even tell your wife your secrets, buddy, even though she's a psychologist and you could talk to her for free. So how are you gonna bring yourself to blab them to some stiff you don't even know? And what, you think payin' somebody's gonna to make their advice more valuable? This is a new low for you, loser."

Rob said, "Let's ask around and get some names."

Finding a therapist was much easier than I thought it would be; *going* was the hard part. But it was instantly helpful when I finally did, because in those first sessions—embarrassed as I was about opening up to someone I didn't know—I not only recognized the shame and guilt I felt for the way I'd treated the people closest to me, Robin and Carry especially, but I also eased up on myself a little, reminded myself that not everything I'd done in my life had been a failure. During the well episode alone, I'd picked a prime spot for the new well, and supplied our house with hand-carried water, and even worked the pump-surge method to success.

When I let myself see that again, when I stepped back and saw the whole of myself, recognized once more that my behavior wasn't an ongoing, singular travesty, but a mixed landscape, it helped to settle me some. And when I recalled the rhythmic hammering of the percussion drill, and the push and pull of the pump-surge method, and the water driven by the ground's pressure up and over the lip of our well and back down again, my own vacillations—the bouncing back and forth between successes and failures, insight and ignorance, light moods and dark—didn't seem quite so alien or destructive.

And it brought to mind something I'd copied down from Leonardo da Vinci's notebooks. He'd written:

Have you ever seen how the water that drips from the severed branches of the vine and falls back upon its roots penetrates these and so rises up anew? Thus it is with the water that falls back into the sea, for it penetrates through the pores of the earth and having returned into the power of its mover, whence it arises anew with violence and descends in its accustomed course, it then returns. Thus adhering together and united in continual revolution it goes moving round backwards and forwards; at times it all rises together with fortuitous movement, at times descends in natural liberty. Thus moving up and down, backwards and forwards, it never rests in quiet either in its course or in its own substance.

My god, when I read the passage again, it seemed as if Leonardo had done a personality profile of me. So maybe I wasn't alone in my pulsing and vacillation. If water moved through the ground that way too, then mightn't the whole world throb with rhythms companion to mine? Maybe that was what had drawn me to this valley seven years before: we marched to the beat of the *same* drummer.

There was hope in that, I had to admit, a hope I needed to reintroduce myself to every so often, especially after the initial months of therapy, when the adrenaline rush of new, dramatic insights was replaced by the slow, slogging labor of deeper introspection, of confronting the voice head on, deliberately, week after week. I'd stand out on the back deck some evenings, look north, and remind myself that even though I seemed to be sinking back or making little progress, there was still hope for growth in me. Because like the water in the ground beneath me, I wasn't fixed or static; I was still moving, always moving. And with luck, over time each movement I made between stupidity and wisdom would erode and deposit knowledge like a tide, redistributing the things I knew and learned in a more useful formation. If it did, then the vacillating process of seeing more clearly, the process of knowing—like hauling water by hand, emptying a large reservoir outside to fill a smaller one inside, *to the well and back, to the well and back, to the well and back—* would become the source of its own progress.

"To see takes time," Georgia O'Keefe said, "like to have a friend takes time." A comforting thought, that: a dream of patience, of water sinking into the landscape, soaking downwards, finding pores of rock in which to rest before rising again. The possibility of an evolving awareness.

Who was to say that wasn't the way things really worked? If our clearest vision, our most refined sensations, didn't come right away, then there had to be some kind of growth involved, some chance for improvement. So what if the progress was slowed by failures now and again; it would also be teased along by occasional wisdom, by pricks of intuition and lucky successes, times when we actually felt ourselves getting somewhere, though we couldn't for the life of us think why or how. It might happen, for instance, when we bobbed to the surface in a swirling pool that had seemed bent on our destruction, and knew the exhilaration of a breath.

So whenever I was on the back deck trying to boost my spirits, reminding myself that retreat was part of progress, I'd recall a triumphant moment or two for balance. In them was the hope for progress in myself.

The moment from the well I remembered best happened on a cold night in February, just after the pipes had frozen. That morning, with Bill Bailey beside me, I'd peeked inside the well for the first time. Now, flashlight in hand, I was coming back to see what I could see, to find out if there was any chance of retrieving water from it. Bill had nicely left his iron bar with me, so I used it to lift the lid off, then turned the flashlight on and looked in. A delicate frost clung like gauze to the upper inside walls of the well and six feet of water stood quietly below. Plenty of water, and it was so clear I could see all the way to the concrete bottom.

It was a frigid night, dark and quiet in the woods. Everything above ground, everything to about three feet underground was frozen. But in the cylinder there was liquid. And the only sound I could hear was the echo of my own breathing, the only smell the richness of earth and water, the only sight the circular well bottom. A simple sensory communion, powerful and unforgettable.

It seems silly now, the unwavering conviction that came upon me then. Silly in light of the distrust I'd held toward our water up until then, the distrust that had convinced me to have a deep well drilled. The distrust that lingers in me even now.

But at that moment I was overswept by a vision, by two things I knew certainly: I knew the water was safe, and I knew I'd scoop it out and carry it inside, even if I had to take it a cup at a time.

Part Three

Air

Chapter 7

The Invisible

It's a beautiful July day, hot and dry, scattered clouds, and while Robin's at work and the boys are with their babysitter—Carry having just turned six last month and Jacob now almost two (though the size of a four-year-old)—I'm canoeing upstream on the Connecticut, thirty miles from home. I'm in search of an old Abenaki village site the spring high-water exposed; archeologists have been excavating it for the last several months, and I want to see what the place looks like before they finish.

The atmosphere on this stretch of river is much different from the stretch near home. Everything seems more open, playful. With the piedmont ridges set well back on either side, the sky feels wider, the terrain beneath it broader. The valley bottom contains lush grasslands, known collectively as the Meadows of the Lower Coös.

Along the banks silver maples wear camouflage across their trunks—thin horizontal shadows, wavering tigerstripes, made by sunlight and rippled water.

On the river itself there's more trickery: along one stretch I notice something brown bobbing in the water well ahead of me, off to the right. A beaver, by the color and shape of its head. But the closer I get, the more I begin to doubt that, since any beaver I've ever seen would have slapped its tail and dived out of sight already. Yet until I'm on top of it, I just can't be sure; it still looks like a beaver floating calmly, eyeing me as I come. At last, I pass within a foot of it and see it for what it really is—the protruding end of a submerged tree limb. And at that moment, as I'm wondering how I could mistake a tree limb for a beaver, there's a loud splash behind me. I turn and see a real beaver fifty yards behind me, swimming out from shore, across my wake.

Hmm.

Finally, there's the maziness of the channel. Between the retreat of the Laurentide ice sheet about fifteen thousand years ago and the organized damming of the river over the last century, the channel has had its way with the easily erodable soils of the meadows, cutting braided courses and looping meanders and, most noticeably, a double oxbow. In his *History of Newbury, Vermont*, Frederic Wells noted that the river had even tinkered with political boundaries:

> The stream has, at several points, worn away acres of land from different farms. It has, moreover, changed its channel in more than one place, and detached portions of land from one town and annexed them to the other, without consulting the authorities of either Vermont or New Hampshire, or the wishes of those who imagined themselves the owners of the soil. An elm which is said to mark the spot where James Woodward in 1762 made his first pitch in Haverhill [New Hampshire], now stands about ten rods from the river on the Newbury side.

The river's tight folds feel like switchbacks on a cliff trail; I'm constantly aware of the reverse-tracking I'm doing. At one point, as I enter a meander, I notice that I'm actually steering south for a bit in order to continue my way north. At first it's frustrating. I'm used to the gentler bends and the straighter travel lines of the river close to home, so the serpentining of this section feels aimless, like a series of detours.

But I'm not going to be able to straighten the river to suit my taste, so the river bends me instead, shaping a different mindset, a broader kind of patience. I stop worrying about the lack of northward progress I'm making, stop trying to push impatiently past each bend as if it were a person in a crowd blocking me from what I wanted to see. Gradually I slow the pace of my thoughts to the rhythm of paddle-pulls: dip in, pull back, lift out, reach forward. With each stroke, water swirls noisily in double eddies behind the paddle, then drips off the flat blade onto the river surface as I bring the paddle forward for the next stroke. After a while the sounds blend into my movements: Dip, pull, *swirl*, lift, *drip*, reach; dip, pull, *swirl*, lift, *drip*, reach. Over and over. I give myself up to it as much as I can, until the anxiety of wandering dissolves and I relax.

Only then do I begin to notice the qualities of the meanders themselves, particularly their mysteriousness, the way they guard the unseen —the way the river ahead, first blocked by the screen of the inside bend, uncurtains itself slowly with each series of strokes. It seems the ideal promise of the invisible.

That the invisible has promise, that it's something substantial, worth considering in itself—and not just, say, a mysterious aspect of groundwater—first dawned on me five years earlier on a fall morning as I was driving on the interstate to work. There was fog that morning, as there often is around here at the start of cold, clear fall days, and it lay like a cotton lid along the valley floor. I was driving on a stretch of road that ran above the fog, open to the clear sky and sun. Rounding a bend I caught a glimpse of rainbow-colored patchwork in the sky—Brian Boland's hot air balloon, floating like a quilted lightbulb above the river a mile to the east. He was taking someone for a dawn ride, I guessed. It was a gorgeous morning to do it.

And I was jealous. I'd been scheming for ways to take a balloon ride with Brian for a couple of years, and when I saw him up there that morning, and the balloon's curved shape so clean against the sky, its colors sharp, I began imagining that I was actually up there too, quietly adrift in the warm light, the soft north wind pushing us through the valley, the rusting hills below us raised like islands out of a whited ocean.

Suddenly the road entered the fog, and visibility closed to within a few feet. I had to brake immediately and concentrate on the white flashes of center line just in front of the car. But even though the visible world had shrunk to a ten-foot radius around me, and I was worried only about steering between the lines, the other experience lingered in my head: the balloon was still in open air above the river and I was with it. So, suddenly I felt in two places at once, struggling inside the fog and floating above it.

This awareness brought both exhilaration and discomfort, an unbalanced sensation of understanding two irreconcilable things simultaneously. Such dizziness is the touch of the mystical, I've read, the paradoxical unknown flowing through us, overwhelming our senses and eventually lifting us, saturated, into light air.

All I wanted to do was tell somebody—swerve down the steep median embankment and up the other side, and intercept cars on the northbound lanes: "Hey! Did you know there was a balloon up there?" I'd say. "Could you even imagine there was one? Of course not. We can't see anything in here. We can barely see each other. But it's there. And I'm up there with it! Isn't that amazing?"

The voice said, "Oh, I'd wanna be there for that. You yelling at innocent motorists. God damn, they'd be so scared, they'd probably try to run you over. I know *I* would."

"Surprise, surprise," I said.

"Well, what would you expect?"

"Nothing from you," I said. "But someone else might understand what I meant."

I guess I dreamed of finding someone who would confirm my sanity. Someone like Black Elk, the Oglala Sioux holy man who experienced his great vision when he was nine; he would have understood my small dizziness, I'm sure. Even at age nine Black Elk recognized the difficulty of translating the unsayable into language. He told writer John Neihardt, "As I lay there thinking of my vision, I could see it all again and feel the meaning with a part of me like a strange power glowing in my body; but when the part of me that talks would try to make words for the meaning, it would be like fog and get away from me. . . . even now I know that more was shown to me than I can tell."

What would there be to tell, if words could shoulder the meaning? For me it would be that the invisible isn't nothing, or the absence of anything; it's an entity, a something that for whatever reason we can't see— like the motes of dust that fill the air in our houses. If we could manage to believe in the invisible, and not confuse it with emptiness, then we might learn how to see it, or at least be alert for the moments when it shows itself, as when a shaft of sunlight beams through the window, exposing the dust in the air.

And if we learned how to see it, I think we'd be amazed by how much of life happens out of sight. In David Bodanis's book *The Secret House*, there are two infrared images: one of a person lying on his back on a bed, the other of the bed after the person has left. Infrared sensors define objects by their heat, and in the first image the reclining human form is clearly visible in pale green. It's the second image, though, that's startling. Even though the body has left—and if we were to walk into the room, we'd swear the bed was empty—the person, as far as the sensor is concerned, is still there. It hasn't the clean lines of the figure in the first image, but there's a shape, a ghostly human shape, lying on the bed.

It sounds strange, but situations like that are more familiar to us than we might first believe. Without giving it a second thought we actually recognize the impression of the invisible every day. For instance, when we sit in a seat just used by someone else, we feel the warmth there and think, "Someone must have just been here." We understand that even though the seat is warm, the warmth doesn't actually belong to the seat; it came from another person. So what we're feeling is the presence of that other person in his or her absence. In this respect we really are aware of the invisible, and though our eyes can't find it, our other senses often detect it easily.

For a minute let's say that's true. Let's say that we're capable of sensing the presence of a person who doesn't seem to be around anymore— who *was* visibly there at one point, but no longer *is*. If we accept that much, we'll have to go one step further and accept that we're naturally capable of understanding a different kind of time too. Because by sensing someone no longer there, we're not only detecting a physical entity, we're also detecting a time, a time outside of this moment, a past enduring into the present. The heat of the person on that seat is also the heat of a previous time, of moments when the person was visibly sitting there.

So the invisible raises questions both about the composition of the physical world and about the possibility of a new kind of time in which moments from the past and present blend together. Like a room whose air is filled with motes of dust, it may be that the present teems with sensory remnants of the past. Part of me thinks that at some level, just beneath our daily conscious functioning, we understand that very well.

Ah, here I go, adrift in metaphysical daydreams again. "River thoughts," Edward Abbey called them, and he was right. There's something about the summertime and the solitude of a canoe and a languid river current that inspires these ideas.

But I can't pass off the invisible as idle thinking. There's more to it than that. I know one reason the idea has stayed with me so long: I can see it at work in my life. As I've thought about it over the last year, since starting therapy, I've begun to recognize how much I've always depended on hiddenness, on invisibility, for survival. For as long as I can remember, my state of mind has been one subterfuge after another, in order to steel my own thin skin and protect my insecurity. "Covert depression," Terrence Real terms it in his book *I Don't Want to Talk about It.* "It is hidden from those around [the depressed person], and it is largely hidden from his own conscious awareness."

That's odd, isn't it? That you could be living with depression, living inside it, grappling with it for years, and not see it, not know you have it? But that's how it was. It was just a familiar, recurrent voice in my head, like a human shape of heat lingering in an abandoned bed. Invisible but present. Very present.

And very invisible too. That I discovered when I began explaining my depression to other people.

"I'm totally surprised," they'd say. "You don't look like someone who has that. I mean, you seem so together. You seem so confident."

"I guess that's the point," I'd answer. "Maybe by hiding it from you, I could hide it from me. If you didn't see it, I didn't have it."

As I paddle around a bend, heading north again, the treeline along either bank falls away and the valley opens like a Chinese fan. There's meadowland on either side—terraced to the west, broad and flat to the east. I notice activity ahead, people swimming in the river and playing on a slant of land that spills out from the bank. While I can't be sure it's the Abenaki site, it wouldn't surprise me—not just because it's supposed to be around here, but because I can now imagine coming upon this spot for the first time and thinking that at least for the warm months there could be no more beautiful place to live. And if the river supported the kind of salmon runs it reportedly did through the early 1800s, I could imagine feeling in this spot a prominent grace.

The Western Abenakis have lived in this region for millennia and were here long before anyone of European descent arrived. Yet as recently as the 1970s, local books proclaimed that European settlers had been the first inhabitants here. It could have seemed that way to those settlers. By the time any Europeans actually lived in this area, the germs they'd brought across the Atlantic with them had long since preceded them inland and decimated native peoples, who hadn't developed a resistance to those germs. Some studies have estimated that the native mortality rate from those epidemics was as high as 90 percent. Regardless of the exact figure, the Abenaki population around here in the mid-1700s would have been far less than it had been in precontact times.

Still, if the Abenaki seemed invisible to the first European settlers in Vermont, it would have had as much to do with the immigrants' nearsighted perspective as it did with the natives' actual population. Europeans brought with them a different idea of "presence," one based on individual land ownership and fixed, measurable property boundaries and stationary agricultural settlement instead of the Abenakis' patterns, which were no less purposeful, but much more fluid, more tied to hunting and gathering, and more cognizant of the rhythms of the landscape and the seasons, the migrations of birds and fish, the movement of deer and elk, the needs of the family collective. But when they were viewed through the conquering European lens, the Abenaki weren't seen and so didn't count. It was as if they'd vanished.

Typical of that perspective is Frederic Wells's description of the Abenaki and their legacy in the Newbury area. He wrote: "Some mounds along the meadows in Haverhill have been thought to be the work of Indian hands. But the few who lingered here after the white men came were degenerate, and soon disappeared."

Though publications have finally begun to recognize the Abenakis' long presence in the region, the same perspective that once questioned their existence still questions the sweep of it. In the common Euro-American view, Abenaki settlement in the upper Connecticut River valley consisted only of two separate bands, each with its own village site — the Sokwakis, whose recognized village was Sokwaki, near Vermont's southern border, and the Cowasucks, who had a village called Cowas (Coös) on the oxbow between Newbury and Haverhill, a hundred miles north of Sokwaki. And that was essentially it, according to the settlers; the land between the villages was an uninhabited vacuum.

Two hermetic living sites a hundred miles apart: it's such a predictable interpretation of Abenaki life by Euro-Americans. But why not? For most English settlers moving north into the wilderness from the more populated areas of Massachusetts and Connecticut, the world was made real by bounded certainties and visual clarity; they began staking out their new farms in Vermont and New Hampshire based on those principles. Of course they saw only two bands living in their contained villages. To have seen anything else would have been to enact a revolution of perception—to have recognized wilderness as burgeoning life instead of an obstruction to settlement, to have seen in murkiness the quality of transparency, to have found in something's remnants the pattern of its whole.

But what Euro-American history has missed by seeing the land so narrowly is the clear evidence of longstanding Abenaki presence, the dozens of known sites in this valley between Massachusetts and Canada, and the numerous others known only to the present-day Abenaki. There's the evidence hidden inside people's private stories and experiences too, things they don't think to share unless they're asked, and sometimes not even then. One Thetford resident told me about finding projectile points along one of the town's brooks when she was a girl; another said his father culled native artifacts from the riverbank near their home; a third mentioned possible sites away from the Connecticut river, on the west side of town. The more I asked, the more I heard, and it occurred to me that if all the fragments of memories were assembled, a clearer portrait of Abenaki presence would emerge, one more faithful to reality than the current one.

I'm not sure the Abenaki would want that, though. While the truth of the native influence in this region, past and present, would open the eyes of the nonnative population, it's hard to imagine what else would be gained by the Abenaki through that revelation. A rewriting of local history? Possibly. New social or legislative changes affirming their existence and rights? Not without some dramatic transformation in our governmental structure. In fact, many people I talked with said the most likely result of any new information would be the result they wanted least: vandalism and theft at newly identified sites.

So, hiddenness has its advantages, I guess. In the context of all that has happened to the Abenaki at the hands of Europeans over the last four hundred years, invisibility provides a kind of safety and freedom that's been hard to achieve otherwise.

When I consider invisibility in those terms, in terms of its benefits, I can't help but wonder about the rightness of what I've done in bringing my depression out in the open. From a therapeutic standpoint, I have to admit it's been helpful. Not curative, but definitely helpful. A lightening of a burden.

But then I think about the ghostly figure in the bed and the warm seat and the motes of dust in the air, and I wonder: should we really see all of that? Is there some good to having the entire spectrum of light accessible to our eyes? Is there some good to having our entire personalities visible too? Would people really want to see me storming around, having an argument with myself? Or should we learn from the Abenaki that sometimes the risks of visibility outweigh the rewards?

I don't know. I see myself chasing after the invisible, talking about trying to recognize it, reasonably sure that if we could, we'd find ourselves face-to-face with the truth of a landscape or the natural world or reality in general, and it would all come clear. We'd know who we were, and where, and why, and we'd know exactly how to be. But then I think about the "truth" of my own personality and imagine having it suddenly and involuntarily and thoroughly exposed. How threatening would that be? How frightening, waiting for the social repercussions? We all know what we as a society have done historically with people who storm around arguing openly with themselves. We do the same thing we do with any undesirable or unsightly societal elements: we put them somewhere out of sight or we willfully unsee them. So when I imagine a world fully visible and then consider my own future in it, I hit a snag. This ideal, transparent reality doesn't sound so wonderful anymore if I'm going to be banished from it. In that case it may be that some things are best left unseen to begin with.

That must have been the kind of experience a local ethnohistorian was working from when I asked him about the Abenaki presence in the valley these days, the physical and spiritual presence.

He said, "It's huge," and then said nothing more.

Huge.

The word lingered around me for a long time, like an echo. I couldn't figure out exactly what it meant, what its dimensions were, until the following spring when the water rose on the Connecticut and the earth fell away from itself, revealing this ancient village site—another remnant of past life, a hidden place that had been there all along, thousands of years away but just inches below our feet.

* * *

By luck the Abenaki site is where I've imagined it, above the spit of land I saw from downriver. As I paddle closer, I realize I have no idea what to expect. I've only seen one other archeological dig, eleven years earlier, when I chaperoned a group of seventh-graders to the Koster site in southern Illinois, where in the 1970s archeologists excavated eleven separate strata of occupation, dating back seven thousand years. Archeology, it seemed to me then, was brutally monotonous work. Lightly scraping dirt away with a trowel or sweeping it aside with a soft brush, then marking, measuring, and recording each small artifact: it seemed nitpicky, attentive labor in a muggy climate more suited to laziness. I admired the crew for their patience.

I don't know what to expect from this crew. I know a group of Native Americans stood vigil at the site before the dig began, protesting what they felt was an act of organized vandalism and desecration by the state of New Hampshire. Would there still be tension? Are the people swimming in the river part of the crew? The protest? Will they be wary of an unexpected presence like mine?

When I reach the spit, which turns out to be a gently sloping cuff of soft sand, a point bar on the inside bend of one of the river's meanders, I find that the commotion isn't a group of protesters at all, or archeologists for that matter, but six canoeloads of boy campers with their counselors; they've been cavorting on the sand and in the water and are starting to pile back into their canoes. After a couple of counselors make a half-hearted search for belongings, the flotilla heads downriver. And the place becomes very quiet.

I climb the bank to the meadow. To my right is a line of cars and vans all facing away from the site, their back ends open, full of equipment. To my left is the excavation area itself—a rectangle about 100 meters long and in some places several feet deep, lying parallel to the river, with one side right along the bank. The rectangle has been divided into smaller metric squares, each one a separate excavation. Since some of the squares have been taken deeper than others, the site has a blocky unevenness, some places terraced like stairs, others dropping abruptly into pits. On the near ground above the site is a line of dirt piles, debris excavated from the site and sifted through screens to catch telltale artifacts. If I remember correctly from the Koster dig, the dirt will be used to fill the area back in after the dig.

An archeological site is a startling example of the scientific mind—
straight lines and square corners laid over shards of the meandering past,
metric quantities given to the knotty contours of others' lives, lives that
were never lived in terms of meters. With its cubic dimensions and fixed
borders, the method seems the product of a mind too limited to ever
really understand the undulations of the life it maps. Yet one paradox
of the Western mind is the fact that sometimes this rigid platform of
reason can, with the help of the imagination, flex like a springboard
and propel itself beyond its rational confinement and into the heart of
that studied life. In other words, sometimes the scientific mind can find
what's most important and least quantifiable about life, even if it does so
by means that some of its proponents would consider "unscientific."

For all of the people working and walking around, the site is remarkably calm. The crew goes about its business quietly and professionally,
though a couple of them, students it seems, banter with good-natured
bravado and take photographs of each other standing in front of their
excavations. From the details of their dialogue—the intimacy and inside jokes—I get the sense of a project in its final stages and a crew that
knows each other very well. Without saying anything I stand well to the
side and watch the scene for a while. There's nothing really dramatic to
see, just the rectangular pit in the ground and the subdued activity inside
it. And there's no specific information to be taken from the site, at least
not to my untrained eye. It will be almost a year before I learn in a newspaper article what the archeologists think they found—a place occupied
by various bands between 1000 A.D. and 1450 A.D.

"What're you doin'?" the voice asks.

"Trying to be quiet," I say. "Inconspicuous."

"Inconspicuous? You're standin' out here in full view! Why don't you
walk around, take a look at the site? You came all this way to see it."

"Nah. I think I'll stay here. Something's not right."

"Whattaya mean somethin's not right?"

"Just what I said. There's a feeling about this place that's unsettled. I
don't know what it is yet, but it's strong." When I first hear those words,
I can't believe they've come out of my mouth. They don't sound like me,
but they are. And it doesn't take more than an instant for me to recognize they're true. Ever since climbing up the bank, I've been uncomfortable. A little shaky. At first I thought it had to do with myself and the
crew, the way I felt like a snoop peeking over their shoulders while they

worked. But the few people who've noticed me here have been very friendly, completely unconcerned by my presence. I suspect they've had a lot of visitors.

So now I'm pretty sure it has to do with myself and this place. And after another minute I recognize the feeling. It's the same one I felt that first year on Cobble Hill, when I came across the secluded field there. It was still and beautiful with unbroken snow, and as I stood there in my snowshoes something told me to leave it be. That's what I'm sensing here, but much more intensely. And then it strikes me: this is a place where the Abenaki once lived, and they've made it clear they want it left alone now. That's what the feeling is—a place that should absolutely be left alone.

"Oh, no," the voice says, frustrated. "You're not gettin' spooked by some Indian mumbo-jumbo."

"It's not mumbo-jumbo," I answer. "It's called respect. You should try a little."

"What's the big deal? The river exposed the site, the landowner called the state, and now they're checkin' it out."

"If the river excavates this place and washes it down to the ocean, that's one thing. That's what a landscape does every now and then, like when the Ompompanoosuc ran into the Ely River and redirected it and left the Ely Wind Gap behind. But when members of an invading culture dig up a former living site of the native culture they virtually annihilated—dig it up against the native culture's wishes, mind you, and then justify it by claiming it'll add to their understanding of native life—that's not the same thing. That's stupid human bullshit is what that is. Just an exercise in power. And it's exactly why the Abenaki don't want anyone to know about these places. If they don't get vandalized in the name of treasure, they just get dug up in the name of science."

"How politically correct of you."

"I'm not trying to be politically correct, okay? It's how I feel. And I'm serious about it, because in the end that's what's going to topple our own culture."

"Whattaya talkin' about?"

"Any culture that doesn't have enough respect for other cultures or for the earth to restrain its own curiosity is doomed."

"Whoa, now we're doomed. It just gets more and more dramatic with you, doesn't it? Listen, what I've been tryin' to ask you is what you think these vandals shoulda done instead?"

"They should have left the place alone. They could see the part of the site that the river opened to view. That should have been enough. They knew what was here. I know what's here. Even you know what's here. Look at this place." Slowly I turn full circle, taking in this dwelling ground snug against a river bend amid a broad, meadowed plain between distant ridge walls. Of course people lived here. How could they not have? "What's the point of digging it up like this, except as a show of disrespect and power? Leave it alone. We don't need to see everything to prove it."

"My, my," he laughed.

"What?"

"You sure have flip-flopped."

"What? I have not. When?"

"About six years ago. Just before Carry was born. Remember? We were on High Peak lookin' at that maple growin' out of another maple and you thought it was floatin' in the air and I was tellin' you that there were definitely roots underneath that root base, even though the whole thing was flipped up and you couldn't see 'em, and you gave me shit because I couldn't actually show 'em to you?"

"Yeah, well that's because you gave me shit for thinking it had roots of air. And you were so sure I was wrong and you were right."

"Wasn't I?"

"I don't know. It's not like you showed me a small glimpse of one live root or something. If you had, I probably would've believed you. But you didn't show me *anything*."

"Neither did you!"

"But the roots I was imagining weren't visible. That was the whole point."

"Then why did mine have to be?"

"I don't know. Because we both assumed they were, I guess. Anyway, this is stupid. Why are we arguing?" I ask him. "What's your point?"

"My point is that if you went up there today and saw that tree for the first time, you wouldn't imagine roots of air. You wouldn't need to. You'd be able to imagine invisible roots in the ground."

"I don't know," I say. It's curious, but part of me thinks he's right.

"Maybe I would. I can't say what I'd do now. But I'm not sure what the difference is, anyway."

"The difference is that little Terry has changed since then. He doesn't need visible proof so much these days, does he? It sounds as if he's sprung his own roots in Vermont, deep enough to feel secure here and have a little faith in the unseen. Isn't that touching."

"Sprung my own roots, huh?" I answer quickly, not at all wanting to give in to him and his condescension. "I don't know. Maybe I have. It'd be nice if that was true. But I'll tell you what: when you start filling my head with your noise, it doesn't matter where I am. It feels like I haven't got a root or an anchor of sanity in the world."

"Thank you," he says, pleased.

One last look around the place and then I turn toward the river. I walk slowly back to the canoe, trying to look relaxed and untroubled, even though I want to run like hell. I still feel shaky. I need to leave this place to itself, leave it far behind. Once on the river, I find myself stroking hard back downstream. I work this way until my arms burn and then I stop, resting the paddle across the gunnels. Sitting still, I let the river take me around the bend.

And in the quiet of that downriver glide something happens—an idea comes slowly alongside, as if washed down from the site as well. With it comes a connection between the past and ourselves, between an archeological site and the unseen, between roots in air and roots in earth. It happens as I float out of the first bend and around the next corner; I begin hearing splashes, and voices talking and yelling excitedly. Through my binoculars, I recognize the group of campers I passed at the site. Perhaps a half-mile ahead, they've landed their canoes near a long rope swing on the Vermont side and are taking turns swinging out into the river from a spot high on the bank. It looks like fun.

My distance from them inspires the kind of disjointed perception of sight and sound we experience at, say, a fireworks display: through my binoculars I watch a camper swing out then drop into the water, sending up a crown of spray. A moment later I hear it: the scream of the swinger, the splashing sound, the shouted response of the other campers. It happens with one camper after another—swing-drop-spray . . . *scream-splash-shouts*—the repetition so consistent that I begin to get used to the rhythm, this two-step perception coming one sense at a time.

In fact, I get so used to it that I find myself waiting for the sound to make the sight complete.

I've learned how to explain the experience mathematically: light travels at 186,000 miles per second, sound at 1,100 feet per second; so when I'm half a mile away, the light from a camper's splash arrives in twenty-six millionths of a second, while the sound takes two and a half seconds. But what really strikes me—more than the separation between sight and sound—is the fact that the event at the bend has *traveled*. It isn't just a case of me sitting in my canoe perceiving something half a mile away; in order for me to perceive it, it has had to travel.

And because it can travel only so fast, there's an inevitable delay, shorter with light than with sound. But in both cases when the event is there it isn't here yet, and when it gets here something else is already there. In other words, my present experience of the campers is their past. A half-mile separation in space means a brief separation in time.

Having just left the archeological site, I think about how much more obvious this connection between space and time is there. I remember reading Stuart Streuver's book about the Koster project and seeing an illustrated cross-section of the different "horizons" or layers of occupation uncovered at the site. The picture shows the eleven known horizons, the last one more than thirty feet below the first and dating back over seven thousand years. What first intrigued me about the picture was the relationship between space and time within the site itself, the idea that each foot deeper into the earth meant a time further in the past.

This connection between deeper observation and receding time is something Timothy Ferris identifies in his book *Coming of Age in the Milky Way*. "The more closely we examine nature the further we are peering back in time," he writes, and then proceeds to subject the back of a human hand to greater and greater magnification until what finally emerge are the subatomic quarks whose structures date back to within seconds of the Big Bang.

But at an archeological site, I imagine something else too, something that happens when you add an archeologist to the picture, along with her own time frame. What happens is that she works in two times simultaneously: her own, which we speak of as moving forward from one minute to the next, and the site's, which recedes from one horizon to the next. So at any given moment, the archeologist's future, her next excavation, will expose a more distant past. Framed this way, no moment in time for her is discrete, solitary. Every new moment is connected to another one, an older one, and she, crouched in her quadrant, trowel in hand, is at the center of it all, time fanning out from her like butterfly's wings.

As I drift downstream toward the campers, the separation between sight and sound narrows and then dissolves until, just across the river from the group, I feel as if I've joined their world, where I hear and see things at the same time, as they happen: a screaming boy swings from a rope, splashes into the water, and his buddies yell their approval. When this happens I think to myself, this connection between space and time, between the present and the past, must go on continually. Why should it just happen here or at an archeological site? Maybe this rope swing and the Abenaki site upstream are like that glimpse of buried root I wanted on High Peak, or like shafts of sunlight illuminating dusty air—brief, dramatic exposures of what's happening around us all the time.

All the time. That's what it comes down to, doesn't it? That the identity of every place—the lay of its land and movement of water, the slant of light and wash of sound—is a product of time as well as space, of what has been, and what is, and probably what will be too. If so, then without knowing it we actually do live comfortably in a number of different time frames at once.

Sometimes the best thing I can do for myself after a trip like this is to come home. There's no canoe solitude or languid current here, and not a shred of time for questions to drive myself crazy with. Carry and Jacob are going full speed and chaos has taken on a life of its own.

Some nights, after Robin and I have spent the last several hours running back and forth between the boys, accommodating their separate needs or refereeing them together; after we've fed them dinner and wrestled on the bed, which either ends up with Jacob in tears, having been taken down too hard by Carry, or Carry in tears, having been taken down too hard by me, or me holding back my own tears, my body bruised from being jumped on over and over; after helping Carry shower and Jacob bathe—simultaneously, but in separate bathrooms, of course, since it would take twice as long to get them cleaned up otherwise and no one would get to bed before midnight; and after helping them get their bedtime snacks and reading to them and singing to them in their separate bedrooms and urging them, *please*, to go to sleep, Rob and I will meet in the kitchen and share a beer and take a breath and try to say something adultlike to each other before passing out.

She'll look into the stove room, where the toy box sits with its top yawning open and most of its contents strewn on the floor, and sigh deeply.

"I just picked that room up before dinner," she'll say. "Now look at it."

I'll answer, "You're making too much work for yourself," which is what I always answer, and which is probably what every husband answers, and which is probably why it falls on deaf ears. "There's no sense in picking up more than once a day."

"But you know I can't stand the mess for that long."

"Then I guess you're doomed to a life of picking up."

"I just want a clean house," she says dreamily. "Just for a little while. I don't think I'll ever live in a clean house again."

"Sure you will," I reassure her with my best husbandly comfort.

"Yeah," she says, sarcastically. "I know. In eighteen years."

"Ohhhh, honey," I answer, with as much mocking concern as I can muster. "There you go exaggerating again. It's only sixteen now."

We bandy time frames around playfully like that, but the truth is we might as well be talking about millennia or milliseconds, so mysterious is the pace of our domestic life. On some nights, four hours go by and it seems as if we've expended four years of life energy. Other nights we stop and look up after turning the page of a book we're reading to the kids and think, My god, where did the years go?

Carry is six. Is that possible? Gone is his interest in the bow and arrow, replaced by action figures and videos and computer games and a growing interest in video games. He wants a Sega Game Gear—one of those hand-held units—in the worst way, his longing so desperate at times that I probably would've done the wrong thing and yielded to it already if it hadn't been for Robin, whose resolve against the idea remains firm. TV was never a big part of her childhood, as it was of mine, and she worries about the effect of so much video stimulation on Carry, about the way he seems drawn to it more than anything else.

To her credit, she has devoted enormous time in steering him in different directions, most notably to drawing and painting, which she loves to do too. Early on it was hard for him; he couldn't always replicate the image he seemed to be carrying in his head, and he'd become so exasperated from time to time that he'd berate himself when he made a mistake and crumple the paper up and throw it on the floor, vowing never to draw again.

"Don't be so hard on yourself, sweet guy," I'd tell him, hearing in my voice an echo of my mother's pleas to me when I was growing up. "This picture's beautiful. And anyway, it's okay to make mistakes. We all do. That's how we learn."

And he has learned, learned how to be comfortable with his imperfection among other things. In fact, he's flourished. His finished pieces all have a colorful, abstract style that's very distinctive and consistent, very much his own. Even more heartening to me, though, is the fact that he's not overly critical of them when he's done. Instead, he's evenhanded and honest: some he likes, some he doesn't. Rob has put as many as possible on display on the walls of the playroom in the basement; the best ones get framed and hung in our mud room, where they greet everyone who enters our house. He doesn't say anything, but we can tell he's proud of that. He should be. They're good. And he worked hard on them.

And as for Jacob, consistent with his settled character from birth, he hasn't had to be steered away from one thing or toward another; he's simply found his own way there, as if there was no other place to go.

That place is music. Neither Rob nor I can pinpoint a specific moment when his interest started; in fact, the longer we watch him the more we think it was another thing he was born with. But we do remember the first time he picked up a stick: instead of using it as a weapon, as Carry had when he was younger, Jacob started strumming it like a guitar.

That was all it took. We bought him a toy guitar and immediately he was serenading us with songs by Raffi and also by Dire Straits, Rob's favorite rock group. Once he showed an interest in the band, she introduced him to their concert videos, and suddenly he became Mark Knopfler, the lead guitarist, and started wearing Knopfler's characteristic sweatbands around wrists and head and mimicking his dramatic kneeling, jumping, guitar-swinging flourishes at the end of songs. It was easy for us to forget Jacob wasn't yet two.

Not so easy for Carry to forget, though. It's been hard on him to see the way things seem to fall in place for Jacob. It was hard enough for him to make room for a sibling, let alone a precocious younger brother.

"I love Jacob so much I'm gonna throw him in the fire," he once told us, verbalizing the ambivalence he felt during the first months of Jacob's life.

But this younger brother came with a precocious love too, a devotion to his older sibling that was so automatic and steady and unconditional that Carry couldn't resist it for long.

We see Carry's goodheartedness—a characteristic *he* was born with, to go with his sensitivity—pushing through his jealousy more and more these days in the form of gifts he makes voluntarily for Jacob. First, he made a holder for Jacob's musical instruments. It wasn't anything more than a big cardboard box he hauled up from the basement, but he and I cut the top flaps off it and made a separate, attached container for the Knopfleresque headbands. Then he spent a long time decorating the box with designs before writing "Jacod's Insterments" in red marker across the front. Jacob called the box his "set," immediately filled it with his instruments, and treated it with reverence.

The second was a work of art, a miniature guitar made from pieces of carefully cut white cardboard: there were two identical pieces in the shape of a guitar body, to form the front soundboard and the back; a rectangular boxlike rib glued in between the front and the back to give the body some depth; a long rectangle glued onto the front to serve as the neck; and another, smaller box-like piece near the bottom to be the bridge. Then he cut out the sound hole in the shape of a heart, cut out another heart shape beside it and wrote Jacob's name on the removed piece, topped the bridge with four foil nobs, where the strings would hook in, and finally colored the neck blue and drizzled multicolored glitter where the pick guard would be.

Jacob wasn't sure what to make of it at first, so awed was he by it and by the generosity behind it. He certainly didn't figure it for a collector's art piece, though, because the first thing he wanted to do was play it. That offended Carry who, like any artist worth his salt, was protective of his masterpiece and suddenly wanted it back if it was going to be misused. Jacob thought Carry's change of heart unfair and started crying, and Jacob's tears made Carry mad, and what had begun as an act of brotherly love had suddenly devolved into raised voices and hurt feelings until Mom the Mediator stepped in and negotiated a settlement that neither of them liked but that saved the artifact: by Parental Decree the Sacred Guitar would be Mounted on a Cardboard Background and Enclosed in a Plastic Frame and Hung in the Premiere Position in Carry's eclectic Gallery in the Mud Room (however, upon polite request it could be taken down by parental hands and held and viewed up close by either the artist or the musician, provided it was done carefully and with no intent to play it).

Disaster averted, at least temporarily—thanks to Rob. All I remember doing during the argument was standing there watching these two young wills colliding—their intentions misunderstood, their signals crossed, everything all garbled together—wishing that I could just turn the clock back one minute, sixty measly seconds, and start again differently with a wisdom I didn't have the first time.

A month after the canoe trip to the Abenaki site, stirred by the thinking I've done since then, I head out into the field again, but this time as part of a group led by local geologist David Laing, who's going to show us glimpses of the valley's geologic past.

According to Laing, the geologic composition of the area owes itself chiefly to the dynamic movement of the earth's tectonic plates. He picks up the story about 570 million years ago, when Earth's land was all one continent, a single mass now referred to as Pangaea I. Growing rifts in the plates broke the great land mass apart, ripping the newly formed North American continent along its east coast. Around here the coastline was much further west than it is now, in central Vermont. The Connecticut River valley didn't even exist.

Actually, it did, just not in the way we've come to know it. As Laing wrote in an article, the separation also left "a series of chunks and slivers of continental crust that had drifted away from the North American coastline on the far side of the rift, where they formed a long chain of low islands that has been called 'Bronsonia.'"

Slowly North America moved toward Bronsonia, colliding with it about 465 million years ago. Among other things the collision helped build the Taconic Mountains on the Vermont–New York border. It also gave us the land on which we, in the upper Connecticut River valley, now live. Below us lies Bronsonia.

"We are island people," Laing says with a smile as we stand on a steep outcrop of metamorphosed volcanic sediments beside the Connecticut River, just below the Wilder Dam, in West Lebanon, New Hampshire.

Thirty-five million years after Bronsonia became part of North America, the northwest coast of Africa knocked against it, pressing the Bronsonian rocks together, rippling them like accordion folds. A hundred and fifty million years after that, all of the continents banged together again to form Pangaea II. And in the 275 million years since then, the continents have split and drifted and rotated to form the configuration we now recognize as the Earth's geography.

At the gorge in Quechee, Vermont—where the Ottauquechee River has cut a narrow canyon in 400-million-year-old quartz-mica schist— Laing tells us that these rocks probably began as an African river delta. During the early stages of Pangaea II, when Africa lay against this part of North America, the river had flowed east to west and deposited the debris that became these rocks.

"A piece of Morocco," someone quips, looking down and the outcrop we're all standing on.

We laugh, but it's partly true: debris from Africa is here.

From Canada too. Later on the tour, as we stand at the base of a sandy bluff in Hartford, Vermont, Laing explains that this bluff may have been an esker, a snaking gravel deposit that was a river in or under the Laurentide glacier. Some of the rocks in the esker, he says, were probably transported from Canada by the ice.

Then he points upward, to the top of the bluff, where horizontal deposits of clay lie like a stack of papers atop the sand. The deposits were the bottom of Lake Hitchcock, the long, spidery glacial lake—extending from Rocky Hill, Connecticut, almost to the Canadian border— that formed in the Connecticut River valley and its tributaries as the Laurentide ice sheet receded. Fourteen thousand years ago, the spot we're standing on was under water, on the floor of Lake Hitchcock.

By the time the trip ends, my head is spinning: Canadian till scattered by a wall of ice about 85,000 years ago, an African delta formed 400 million years ago, a volcanic island chain that became our home valley 35 million years before that. It takes weeks before I get some perspective on it all and fit it in with everything else I've been thinking. But gradually I understand that the rocks confirmed what the Abenaki site exposed: reality is both laminate and interwoven. Just as one village site was laid over another and another and another, like newly grown layers of skin, the rock formations beneath them had been built and stacked too. Around here they'd also been folded and cracked and pushed over like a deck of playing cards, but regardless, the result is the same: a stratified collage of time, an endless present that seems to wrap itself continuously around the ever-accreting past like a shawl.

But the present and the past aren't alone here. The future's sheathed together with them too, inseparably. Almost identically. When I look north up the valley from home, I can see all the time frames fit together now, because what I'm seeing is not only the Connecticut River valley's 465-million-year past; it's also Bronsonia's 465-million-year future.

When I look north up the valley from home, what I begin to see is the kaleidoscopic character of this place. Of any place. The face of this valley bowl — as I glimpsed most clearly several years ago from the ravine on the Vermont piedmont, where I sat out the thunderstorm — is defined by two parallel ridgelines and the line of water that has carved out the space between them. But underneath, those features owe themselves to a volcanic island chain and an African river, Canadian ice and a glacial lake, which all commingled over time to form the present place I now call home. The valley's essence, its identity, it seems, is an incredibly complex, freely moving mosaic, slowly rearranging itself even today.

The prospect of the landscape not being fixed in time and space, but constantly churning instead, should scare me to death, shouldn't it? Me, who usually feels a stab of panic when the surroundings aren't what I'd anticipated, who can turn angry or sullen when events change unexpectedly, who can't sustain a mood for longer than a day, a minute, a second. But it doesn't. Isn't that strange? Maybe the voice was right; maybe I have changed. But, wait. Of course I have. That's the whole point: I'm supposed to change. We all are. Rob, Carry, Jacob, Griffy. We change and vacillate and misstep and repeat and with any luck redirect ourselves. Last year, with the help of the man with the map and our shallow well and frozen pipes I learned about the continuous countermotion of groundwater through the earth, its uneven pulse that my own behavior and thinking seemed to mimic. I wondered then whether the whole world might pulse that way too.

And now I see. After my journey to the Abenaki site and the geological tour with David Laing, I see it's true: *nothing* here stands still—not water, not land, not air, nor anything created from them. Everything is moving. The core of this place is not really a single geologic feature like the Palisades' granitic dome, or a specific geographic location like Thetford, or a particular person like Asa Burton, but a collective movement, a long-standing, ongoing transformation through which everything pulses.

This must be the quality Gary Snyder calls *fluidity*:

> Even a 'place' has a kind of fluidity: it passes through space and time. . . . A place will have been grasslands, then conifers, then beech and elm. It will have been half riverbed, it will have been scratched and plowed by ice. And then it will be cultivated, paved, sprayed, dammed, graded, built up. But each is only for a while, and that will be just another set of lines on the palimpsest. The whole earth is a great tablet holding the multiple overlaid new and ancient traces of the swirl of forces.

The multiple overlaid new and ancient traces of the swirl of forces: such an evocative phrase, so full of motion sounds. It seems finally like the answer I've been searching for all of these years—the key to the valley's character, and to my own. *Multiple. Overlaid. Swirl of forces.* And it captures best in both sound and meaning what I sense when I stand on the deck and look north up the valley now, up that familiar valley. It's as if I'm looking through a window onto the world that's also a mirror into myself. I am home.

I'm never more aware of how complex that home is than when I bump into the invisible. It doesn't happen often, but just enough to keep me expectant.

It happened most recently on an autumn morning when I was driving past the field beside our house. Looking out of the car window I saw a sight that certainly must have been there before, though I'd missed it until then: strung vertically between the tall grass stalks were dew-soaked spider webs—thousands of them across the acres. Knee-high, circular, and a foot or so in diameter, they looked like high-tech eyes or little satellite dishes. They looked like the wagon-wheel pattern I'd placed on the valley years before, strung with my spoked lines of travel. They looked like swirled beadwork raised bright against the straw-colored backdrop of the hillside.

Later that morning, when I drove by the field again, they were gone. I stopped the car, got out, and stood at the edge of the field. As hard as I tried, I couldn't see them, even though I knew they were still there. I mean, the spiders hadn't packed them up and moved on, like a traveling circus. All that had happened was that the dew had dried. But it was as if they'd evaporated.

Beside that field I was overrun by a twin awe: first, that such intricate and abundant beauty should exist in a place that had always seemed so empty; and second, that it should be so damn hard to see, even though I knew exactly where to look.

The first awe was my awareness of fluidity, I think. I had a dizzy sense of being surrounded by, and a part of, the unseen whorl of life. There was a taste of the mystical in it too, something broader than myself—"a more permanent self," as Louis Dupre writes, "in which space and time are transformed into vistas of an inner realm with its own rhythms and perspectives."

The second awe was the grip of mortality—of an existence closely hemmed on all sides, of a mind easily fooled by the invisible and eager to distill the world into a manageable pattern of lines and angles, or semi-circles and radii, or a wagon wheel split in half, or any other interpretation it can rationalize enough to call "reality." This is the weakness in my psychological system, the tendency the voice inspires and capitalizes on, the awe the depression feeds on. I may feel fluid sometimes, but at other times that fluidity seems like a hoax, because overriding any feeling of permanence is the fact that I'm mortal. Irrevocably, irrepressibly mortal. I will die. Life, or my awareness of it, will end. And in my least secure moments, when the depression has set in and I'm feeling alone and vulnerable, I can think of little else but how silly it seems to be born into this world, and then grow up and mature and feel the tingle of fluidity from time to time, only to have to die.

I wonder if being overlaid by the swirl of forces means struggling with those two awes. Do other humans struggle with them? I have no idea. All I know is that I do. And when I do, it's as if I've joined a tug-of-war already in progress between mortality, with its inherent insecurity and confinement, and the invisible, with its fluid permanence. I find myself slipping into a familiar tension between extremes, a duel of counter-motions. I pull myself one way, then get pulled back the other.

There's no question which way I want to pull: now that I've heard the whisper of the invisible and seen that it isn't nothing, or even something, but *everything*, that it fills our lives the way orb webs fill a field, I want that. I want fluidity, the swirl of forces, the deeper self; I want to be conscious of it all now and forever.

Sometimes I can actually feel the fluidity on my side. Sometimes, when I'm worried about missing out on life as it really is, or when I'm feeling lonely, the way I imagine it must feel to be dead, I can calm myself by remembering what the water and the land and the air have taught me.

"Everything comes in its own time," I tell myself. "Just remember the meandering river and the folded rock and the plowing glacier. Remember the promise of sound after a flash of light. And be patient."

There are days I can be patient. Breezy days when I'm unhurried and needless, content not to push on and make myself see someplace else, learn something new. Then I'll pull my paddle from the water and drift slowly downstream on the Connecticut. Or beside a secluded pond on Potato Hill I'll lie in the sun longer than usual and think of Robin and the boys, feel the joy of returning home to their hugs and body slams. On those days the voice is absent or quiet or affable.

I'm able to string more of those days together now, thanks largely to my therapy. When Rob and I look back, we can see a definite rhythm to my moods during recent years, a six-month separation between depressive troughs. I know that half of that time, though, was spent descending into the depression and recovering from it, so there were perhaps three months out of every six in which I might have felt well enough to experience a handful of patient days.

On those days, now as then, I can come home and be totally present, have energy for the boys and look Rob in the eye, talk with her and touch her and hold her.

And if she's feeling safe enough, she might say, "I can't believe it took me so long to see your depression. I mean, it was right there in front of me the whole time, and I didn't see it."

"But you *did* see it," I tell her. "You were the one who picked up on it first. And you tried to tell me. It just took me a while to hear you."

"But still, it didn't click inside me for so long. And it's part of my profession. I see it a lot in my work. I know the symptoms, I'm supposed to be able to recognize them. But here I was, living beside them every day, and I didn't."

"I don't know. I think you're blaming yourself for something you couldn't possibly have understood until it had actually happened to you. Seeing depression as a clinician has got to be different from seeing it as a spouse." I mean this sincerely and think she's being irrationally self-critical. But when I read Terrence Real's similar lament in *I Don't Want to Talk about It*, I begin to understand her reaction better.

After explaining that his decision to become a therapist was influenced by his unconscious desire to understand his father's chronic depression, Real writes:

> One might think that I would have brought to my work a particular sensitivity to issues of depression in men, but at first I did not. . . . I was not prepared, by training or experience, to reach so deep into a man's inner pain—to hold and confront him there. Faced with men's hidden fragility, I had been tacitly schooled, like most therapists—indeed, like most people in our culture—to protect them. I had also been taught that depression was predominantly a woman's disease, that the rate of depression was somewhere between two to four times higher for women than it was for men. When I first began my clinical practice, I had faith in the simplicity of such figures, but twenty years of work with men and their families has lead me to believe that the real story concerning this disorder is far more complex.

"Maybe," Robin says, "I'm just worried that if I didn't see the symptoms right away before, I won't pick them up early enough the next time. If there is a next time."

And there is. There always is. No matter how content I feel on those fluid, patient days, or how loved, and no matter how many of those days there are, they'll end, and after a while something will happen to make me insecure—something I'm never prepared for—and the hostile voice will return and pick on my insecurity and gain power over me, and then I'll feel the renewed tug of mortality, of need and want and impatience and loneliness. Feeling the specter of mortality crowding me, I'll finally be overcome with worry. And as I, resisting all the while, get dragged toward the voice, my heels gouging troughs in the ground, life becomes shadowy and stunted again, the landscape far off and untouchable. And the voice and I reenact the kind of confrontation we've had since I was a boy, a confrontation he knows how to win if he wants to.

"Why'd you come back?" I ask, maybe too defensively.

"Because you're stuck here alone and you need me," he answers.

"The hell I do. There's no room for you here."

"Sure there is," he says calmly. "Plenty of room."

"Not any more," I say, trying to sound strong and confident. "I'm all filled up with everything I've seen these last eight years. The world is so beautiful, and complex, and dynamic. And I have more in common with it than I ever thought. That's what the valley was telling me when I first saw it." As I finish, I think for a split second that I might actually be able to resist him this time.

"Oh, please. There you go, gettin' all romantic on me with your talkin' valleys again. Spare me, will you?"

"But I'm serious. I feel like I understand the world and myself better than I ever have. Why is that a problem for you? Why do you step in and block me whenever I start to see things more deeply or clearly? Why don't you just get lost?"

"You tell me. If this place is so gorgeous and fluid, and that's what the world's all about, and if you can see it now, then why don't you just keep on doin' it? Go ahead. See it all the time." This is how he takes control of our exchange, testing my fragile confidence, forcing me to the edge of it.

"I can't," I say quietly. "You know I can't." And not feeling secure enough to add, "No human can," I have no other defense to offer—just this "can't," this confessed insufficiency that gets added to all the others I'm feeling. And that quickly, my resolve erodes; I'm on my way down.

He knows it. "Aw, why not?" he asks, condescending.

"I don't know. I wish I *could* see everything all the time. I hope there *is* some reason you try to keep me from doing that. A *real* reason. Not the kind you make up just to justify yourself. If there is, I'll figure it out."

"Oh, like the way you've figured out the valley?"

"Yeah, like the way I've figured out the valley."

"Like the way you see your personality reflected in it?"

"Not quite that simple, but something like that, yeah."

He almost spits with laughter. "I guess I'm safe, then."

"If you say so."

"But what if there *is* no reason, except that that's the way things are supposed to be? What if I'm woven into you the way all the pieces of your fairy-tale landscape are woven together in ecological harmony? What if you can't get rid of me, no matter how hard you try?"

It's a brilliant question, because it uses my own hard-won understanding of the valley to corner me in my worst nightmare. "I don't even want to think about that," I say.

He's almost got me, so he tugs harder. "Nah, c'mon. Give it a shot. What if I'm here simply because you're here? And what if I'm scheduled to stay as long as you're alive?"

He's dragging me toward a precipice I don't want to look over, toward a place, a future of utter hopelessness, where I'll be at his beck and call forever. God, I don't want that. I don't want to have to listen to him for the rest of my life. These isolated episodes of depression, though recurrent, are excruciating enough as they are, damaging enough to Robin and the boys too. A future of continuous depression is unthinkable, so as he pulls me toward that brink, threatens me with that future, I imagine the most drastic alternative I can.

I say, "Then I'd have to think about killing us both." I mean it to sound grim and committed, but it's too late for that. It comes out scared instead.

"Oooo, kill us both. Sounds excitin'. Too bad we both know you don't have the guts to do it. You're too afraid of dyin' yourself."

He's right and that's the worst thing of all. Worse than the threat of living under his power forever. Worse than the thought of killing myself in order to get rid of him. Worse than the realization that I could *never* kill myself. He's put me between two alternatives—him or death— knowing full well I don't like either of them. So all I can do now is admit that he's right and reach out for the only solid thing I'm sure of right now—which, ironically, is him. That's the worst moment of all: feeling so defeated and scared that 8I reach for him.

"I know I am," I say, hating the sound of my own voice. "I know." And with that repeated concession, I surrender. I have no mind of my own anymore. We're on the same wavelength now, the same side. His.

"Take it easy, okay?" he says, his tone softening. He knows he's got me. "We've been together a long time. I'm not gonna hurt you. I'm not gonna make you think about killin' yourself. That wouldn't do either of us any good, would it?"

"No, I guess it wouldn't," I admit. And instead of hating him, as I have until that moment—hating him for goading me here, for taking advantage of my vulnerability, turning my eyes toward darkness; instead of dreaming of his demise, as I sometimes do after a depressive episode, of how I want to bury him in the backfilled hole of some desecrated site; instead, I think how right he has been about me and my life, how much he really cares about me and needs me around. I can't believe I haven't ever seen it before. I can't figure out why I haven't listened to the voice more often. I mean, *really* listened to him. I know I should stop resisting him and hear what he has to say, because he usually makes sense if I give him time to make his point.

He finishes pulling me in and puts his arm around me.

"There you go," he says gently. "Now you got it. See? You should be thankin' me, not cursin' me the way you do, because I'm here for your own good. The talkin' valley isn't. The fluid world isn't. You know why I know that? Because one day this world is gonna end, whether anyone likes it or not. One day the sun is gonna lose steam and wink out, closin' like a bloodshot eye for good. On that day the only grace you'll have will be eight minutes. Eight minutes given to you by your blindness and the limited speed of light. Eight minutes of little daffodil ignorance before the last light arrives from the extinguished sun, before the sky goes dark and you—so alone, so surprised—begin precipitously to freeze. Why would anyone trust a world like that?"

Chapter 8

The Voices

I

He's in control tonight. And impatient. He wants someone to fight with, someone besides me—a *real* enemy, who offers some resistance, who even fights back, who isn't so afraid of her worthless mortality, who isn't so easily convinced that she's just a sack of dilute minerals bound alone for frowning, ignominious darkness and death and decay and damnation. He wants the pleasure of bringing her down, making her miserable. On this night, like most others when he's feeling this way, he works up his anger as he waits for her.

"Where the hell is she?" he asks me as I scrub out pans at the kitchen sink. "She's five minutes late."

"At work, I guess," I say meekly.

"What, like she doesn't know you did the shopping this mornin' and have been with the kids since four and cooked dinner too and never got a chance to write?" he shouts. "What if you had somethin' to do now? What if you had somewhere to go?"

"But I don't."

"But what if you *did*?"

"Then she'd be here," I say. "C'mon, you know she's not trying be late. Anyway, she's the one who makes most of the money around here. She's got to work. So give her a break. She'll be here."

All the time I'm staging this little stand on her behalf, I know it's useless. The voice doesn't need or want to be rational tonight. He simply wants a target.

"So you say. All I know is that she told you she'd be home at six-thirty. So give *me* a break."

A ninety-acre oval sunk into the broad valley plain and surrounded by open fields, Conant Swamp lies west of the Connecticut, a mile and a half north of our house. No matter where you look at it from—whether from the east on Route 5 or from Latham Road on the west—it seems pretty unremarkable, just another island of woods bristling out of the cropped farmland.

But it's different. More than any other spot in the valley, it pulled me toward it. Looking back now, I can see that our meeting was inevitable, but I didn't understand that then, didn't understand what the swamp held for me until I finally explored it. And in the four visits I made to it over a period of seven months, it challenged me as much as any other place I wandered though, challenged me physically and mentally, challenged me to walk into it and get lost and learn its natural history and listen to its voices—sounds and memories of death and love and depression and relief—and make a place for all of them within myself. And then it challenged me to find my way out again if I could.

Though it might not seem so, ninety acres is plenty of space for that kind of challenge. Certainly, Conant Swamp is small compared to other swamps in the state, but it's reasonably big for our area—much bigger, for instance, than the peeper swamp across the street from us on Cobble Hill. It's also the only one at the bottom of this valley bowl. That doesn't mean it's the only wetland; there are a number of marshes alongside the Connecticut, riparian wetlands whose size and condition fluctuate with the level of the river. But what makes Conant Swamp interesting is that it sits off on its own—a half-mile away from the Connecticut and also about fifty feet above it in surface elevation. To get there from the river, you have to climb over a steep hundred-foot bank to the west before following the gradual tilt of the valley plain down into the depressed pocket where it lies.

This pocket is, I suspect, one of two reasons the swamp retains moisture: the concavity of the landscape acts as a sink, collecting all of the surface runoff from the higher ground around it, especially from Hough-ton Hill and High Peak to its west. The second reason is less obvious—in fact, it's invisible—and completely my own hunch: it's the Ammo-noosuc thrust fault, a west-dipping, Mesozoic crack in the land that runs directly underneath the swamp and that may—through even the slightest, lingering distortion of bedrock—give groundwater access to the surface. If that's true, then water is drawn to this same spot from both above ground and below.

I was drawn to the place too, from the moment I first heard about it. Isolation was the primary allure, to be sure. Not the remote isolation I've felt in some spots in the valley, but something more curious: an unguarded kind of secrecy. I was intrigued by the way the swamp seemed to stay so hidden, though clearly visible in the middle of the valley. To do that, a place must be forbidding, I thought—thick, tangled, mucky, haunted. A place like that might get someone irretrievably lost even though he was within shouting distance of the nearest house.

I asked Arthur, our town lister at the time, about the swamp, any interesting stories he'd heard. He couldn't think of any, except for ones related to hunting—who'd shot what and when. He recited a string of different families who'd owned it, or parts of it, over the years, but that was all he knew.

I guess I wasn't surprised to hear how often it had changed hands. Until recently wetlands in America have been considered the lowest form of wilderness, with no intrinsic purpose or value. As William J. Mitsch and James G. Gosselink write in their textbook *Wetlands*, "Prior to the mid-1970s, the drainage and destruction of wetlands were accepted practices in the United States and were even encouraged by specific gov-ernment policies." According to a table in their book, Vermont lost more than a third of its wetlands between 1780 and the mid-1980s. Given that, I suppose it's to the various owners' credit that Conant Swamp still exists.

Still, I was disappointed by the lack of folklore. The Conant Swamp I was imagining ought to have had a lurid history, a Headless-Horseman kind of past, stories with sedate beginnings, crooked middles, and no discernible ends. But if there were any like that, Arthur would have known them, or at least heard of one.

Anyway, that didn't stop me from wanting to go in there.

One day I asked Bob, a friend of mine, whether I could use his property as a way into the swamp.

"You going in there alone?" he answered.

"Yeah, I was planning on it," I said. Then, when he didn't react, I added, "Idle curiosity. . . . Y'know."

That was half true. Curiosity, yes. Idle, no. As I've said, I was drawn to the swamp from the start, though I didn't know why at first. I think there was something mysterious and paradoxical about it that reminded me of my first view of the valley nine years before. And I had a hunch. A hunch that I'd find something in there. I didn't know what it would be—maybe something macabre, like a lost child or a dead body. Or maybe something that would confuse me more—a glimpse beyond the invisible, beyond fluidity and the swirl of forces. But it would be something, I was sure. Something important. Something you could hold on to, even if you didn't want to.

Eyebrows arched, Bob tilted his head to one side but didn't say anything.

"What?" I asked.

He told me he'd only gone in there once. He was with his boys, chasing a raccoon they never caught, because the swamp had sucked him in up to his hips and damn near swallowed him whole; it took him two hours to get out.

"Just let me know when you're heading in," he said.

"How come?"

He smiled. "So if I don't see you for a while, I'll know where to send the Fast Squad."

The first time I saw a dead body was during my second year in college. The older brother of one of my roommates had been killed in a single-car accident, and though most of us had never met him, we all drove to the wake to support our grieving friend.

The room at the funeral home was crowded, the casket against the far wall, open. After an awkward conversation with our roommate and his family, we stood apart, clustered uncomfortably, wondering what to do next. It seemed most respectful to do what we saw others doing, which was to kneel and pray in front of the casket.

I only looked briefly at the brother's face before kneeling: it seemed a touch yellowed and powdered but reasonably natural, not too much like a mannequin. But as I knelt, I was jarred by an internal quaking—the rumble of land masses thrown together, the most severe quaking I'd ever known. In my ears a concussive ringing, the kind that comes after your head's been slapped. I couldn't hear a thing but my ears themselves. So I was hunched on my knees, eyes closed, in this quaking, ringing world no bigger than myself—a world, I now think, of profound terror—praying into the side of the casket, saying something like, "Excuse me, I never knew you, but . . . ," and wondering whether my legs would work when I pulled myself to my feet.

I rose feebly, and then a couple of us said goodbye, stumbling away from the casket and out of the room before stopping in the lobby to wait for the others. To our left was the front room, brightened by large windows, empty except for another casket open at the far end. I don't know what inspired us to do it, but we crept over to the casket to peek in. Laid out there was a boyish man about our own age. Though his hands were arranged neatly on his stomach and his expression was calm, he looked only partly human. Across the lower half of his bloated face was a bluish shadow the makeup couldn't conceal. And I suddenly imagined violence done, self-obliterating violence, a shotgun under the chin, the remnants of which the undertaker had used to piece the face together again.

After a minute we all looked up, caught each other's horror, and then did our best not to laugh.

We stood there in silence.

Finally, I turned to the others. "Now, *that* guy," I said, "is *dead*."

2

By the time Robin gets home, he's made himself livid.

"Hi, sweetie," she says to him as she walks into the kitchen. "How was your day?"

He grunts but doesn't turn around, stays facing the sink, scrubbing. She's ten minutes late.

Moving away warily, she greets the kids, hugs them, asks them about their day, then changes out of her work clothes. Later she comes back.

"Ter?" she says quietly. "Everything okay?"

Sliding away against the counter, as far away as he can get, he half-turns toward her without making eye contact. And he waits. Distance and accusing silence are his opening weapons.

"Ter?" she asks again. She's barely gotten home, but she already sounds hesitant. She knows this routine. "Are you angry about something?"

Still only half-facing her, he says, "No, I'm fine," in an innocent, sing-songy way they both know means something else, means he's definitely not fine, that something is very much wrong, and that she'd better brace herself for what's coming, whatever it is.

Starting from Bob's house, I went into Conant Swamp on skis the first time, on a bright Sunday morning in February, and traced its length—back, forth, and back—in three parallel paths. Two things struck me that morning. First, the swamp was incredibly easy to get around in. There was the usual tangle of roots and stems, but nothing sneaky, no surprises. The snowcrust bore my weight like a sturdy platform. Second, I never saw a living creature. I mean, I thought I'd see *something*, even the blur of something dashing for cover. But no luck. Heard some chickadees, saw a web of tracks—crow, rabbit, weasel—and that was all. After everything I'd imagined about the swamp, it was strange to find the place so light and deserted, so quiet, so dead. I couldn't believe that that was the way it really was.

I didn't get back there for several months but finally returned on a late-summer afternoon. I was short on time that day, with only a couple of hours to spare before I had to head home to take over for our baby-sitter, so I was hoping to make a hurried reconnoiter, to scout the place out for future trips.

Behind Bob's house the whole terrain looked different, everything clogged in vertical green instead of the flat white that had held me so easily on skis. Bordering his back field was a tall hedge, and once through that, I found myself in a glade of long, drooping marshgrass, a couple of hundred yards west of the swamp.

I waded into the grass, taking one awkward step, then another. The third one was a long one. Down. I toppled over into a canal that had been shrouded by the grass. For a minute I lay against the far bank, startled, my left side sunk up to my waist in muck.

Griffy looked down at me from the other bank, ears up.

"Don't even *think* about laughing," I said to him.

As startled and embarrassed as I felt, there was one advantage to being down where I was: I could actually see a tunneled path ahead of me, where the canal pushed through the arched grass. If I'd been adventurous enough to slog through the muck, and had had enough time, I could've made a pretty direct line to the swamp, I think. But the plunge had seemed such an inauspicious start to what was going to be a hurried, abbreviated trip anyway that I decided to honor the omen: this trip was over. So, after pulling myself from the canal, I called it a day and went home.

"So," a psychiatrist asked me during an introductory visit, back when I was first looking for a therapist, "what do you do for fun?"

I waited, hoping something might come. Maybe something from the ceiling, like a feather floating down, some happy answer that would land in my lap. Nothing did. Instead, there was just a thin, high noise in the air, a hissing from the stuffed leather chair as I shifted in it.

Fun. What a question.

"Man," the voice said to me, "he just doesn't get it, does he, because if he did, he wouldn't have asked such a stupid-ass question. What a quack. And you're gonna take his advice? Uh-uh. I wouldn't. Shit, I can tell from here he doesn't have a clue."

Fun. Did he really want to know how fun it was inside me? What it was like to live with this voice that was cutting him down as he sat there waiting for my response? If I'd had it with me, I'd have given him the best description I'd ever read of my inner neighborhood and all the fun it held. It was written by physicist Roger Jones in his book *Physics for the Rest of Us*:

> If you could beam yourself into some typical average locale in the cosmos, you would be millions of light years from the nearest source of light. You would, in fact, be surrounded by an oppressive inky blackness, pierced only here and there by the tiniest pinpoints of light which would be the images not of stars, but of galaxies that are so far away as to be all but invisible.

I shifted again in the hissing chair.

"Fun?" I said and struck a casual, thinking pose. "Hm. I guess I just don't see things that way."

3

Far enough into the depression and it starts to feel as if I'm being held hostage in a basement room—thick-walled, windowless, one white porcelain light fixture broken and dangling from the ceiling. The voice knows that all he has to do is hold me here, silent, long enough for me to forget that there's any other place but this; if he can, then he might be able to rule the roost for weeks, months, before I realize where I really am. But in those first days, if I can just manage to speak, to name this place, hear myself admit how I'm feeling—tell Robin, "Yeah, I guess you're right. I guess I'm depressed."—then suddenly the walls thin, the light fixture works, a door's outline appears in the far corner, and I can see a way out, an escape on the back of my own words.

After dinner, while helping Rob put the boys to bed, he's careful not to talk to her. When he's around the boys, he's lively and entertaining—wrestling with them, telling jokes, reading their nighttime book in funny voices. But when she comes near, he barely even acknowledges her, except to answer questions.

"Did Carry brush his teeth?" she asks him.

"Uh-huh."

"And Jacob's clothes, do you know where they are?"

"There," he mutters, annoyed, pointing to the railing at the top of the stairs, where the clothes are stacked.

As she passes him in the hall, her shoulder almost touching his, he looks away.

Stopping, she turns around quickly. "Ter," she says. It comes out as part plea, part exasperation. It tells him she knows he's making a point even though she doesn't know what it is, and if he has something to say, he should just say it, instead of dragging it out and making her wait and guess.

"What," he answers, raising his voice. "What's your problem?"

The playful banter coming from Carry's room stops. Six and a half years old, Carry knows the sound of an approaching storm; he's heard this voice before. Jacob, at age two, hadn't picked up on anything, but he goes silent when Carry does. Now they wait too.

Robin won't take the bait; she will not argue tonight. She only sighs, deeply, and as she does, her whole body seems to drop a little, shrink into itself, like a cake just pulled from the oven.

"C'mon, do we have to do this?" she asks. Then she shakes her head, and picking Jacob's clothes up off the banister, goes to put them away.

After I fell in the canal, I knew there had to be an easier way into the swamp. The next day I drove right up to it from the east side, through fields I'd gotten permission to cross. Once parked, I assembled my gear and put on my green rubber K-Mart chest waders. I hadn't used them in years; they were heavy and about as flexible as a suit of armor, but I knew they'd protect me from any surprise canals.

"Whoa. Lookin' good," the voice said sarcastically.

"Feeling good," I said.

"No, I mean *real* good," he said.

"That good, huh?"

"Real, *real* good," he said.

"Okay. Enough. I get it."

Outside the treeline on the northeast corner, Griffy and I started through a wide, crescent border of jewelweed, chin-high and tangled. Walking through it in those waders was like trying to comb snarled hair with fingers: it was all yank and snag and yank, and by the time I got through, my thighs were burning and I was out of breath. I might have turned around and walked right back to the car if it hadn't meant going through that damned jewelweed again.

And anyway, I was in.

Inside the trees, the climate was different. Out by the car it had been an autumnal day—bright sun, scattered clouds, steady wind out of the northwest, cool, dry. But it was warm twilight where I stood now. Humid and still. Sort of mysterious, the way I'd imagined it might be.

So there I was, finally inside the swamp and ready to undertake the Plan I'd hatched: first, I was going walk the perimeter of the area, staying just inside the treeline, and come full circle if possible, finishing right where I'd started—in this case, by the jewelweed. That would give me a broad sense of the terrain, I hoped. Or at least of its rim. After that, I'd very deliberately explore the interior: starting at the east side of the swamp, I'd walk back and forth between the north and south ends, trying to keep straight, parallel paths spaced apart by about a hundred feet, until I got all the way to the west side. Once I was done with the perimeter and the interior, I'd use all the notes I'd taken to draw my own map, one that would combine the separate pieces of swamp I'd walked through into a single layout I could see all at once. That would go a long way toward getting to know the place, I thought, and give me something of my own making to hold.

If the Plan seemed contaminated by some of the straight-line thinking I'd criticized at the Abenaki dig site the summer before, I had to admit it was true: I was guilty of the hypocrisy. But only to an extent, because what I was really trying to do was follow the eclectic methodology of Ken Bannister, the geologist I'd talked to years earlier when I was studying groundwater and wells. I remembered the straight lines he said he would walk across a potential well site, first with his magnetometer and then with his VLF receiver, and the dowsing sense and intuition he'd use to help him make his final decision. I remembered the colorful clutter of data plotted on his master map. That mixture of methods, that mosaic of thinking—which seemed to give the well site the chance to determine itself—was the kind I most admired, the kind I wanted to emulate in Conant Swamp.

But no one was going to mistake my methods for Ken's. I hadn't had his background or training; I hadn't had *any* background or training. Or nothing formal. Only what I'd picked up from my own experiences over the last nine years. So the perimeter I was planning might have the loose geometric shape of a circle, but it probably wasn't going to be symmetrical, because I was going to let it be guided by the vagaries of the swamp border. And the lines through the interior? I hoped they'd be straight, but I wouldn't have any way of guaranteeing it apart from the headings I took from my compass, and no way of standardizing the spaces between each line except for counting out the same number of paces as I went from one to the next. No, there'd be a good deal of human error involved in my Plan—and with any luck, there'd be intuition and patience and faith too. I hoped that would make for a good enough map. Good enough for me, at least.

Aligning myself and my compass on the east edge of the swamp, I started off.

Heading south I took advantage of a deer path snaking along the gently curved eastern border. What I couldn't believe, as I tried to stay on the path, was how something as tall as a deer could get so small without shapeshifting. There were places—tunnels between bushes and under fallen trees—even Griffy had trouble squeezing through. Often I had to drop down on all fours and crawl. Here and there I'd happen by a spot where the long grass had been flattened into an oval, and I began to prepare myself for the surprise of a doe and her fawns springing up in front of me.

"I wouldn't worry too much about that," the voice said. "There isn't an animal within a mile of here."

"How do *you* know?" I snapped.

"Oh, gee, let's see. Maybe because you're makin' more noise than Sasquatch, you rubberized footstompin' brush hog."

"It's that loud, huh," I said.

"Brutal. You might as well be shoutin'," he said. "Even the deafest deer is long gone."

From then on I did my best to walk more lightly. By the time I reached the south end, the terrain had dried and the ground cover had thinned. The air was brighter, too: hemlock, beech, oak, and white birch stood tall and widely scattered. But that didn't last long; several hundred feet later, at the southwest corner, the terrain transformed again. I came up against the edge of a swampy area, infused with an intense gray-green light. Yellow birch, sugar maple and hop hornbeam were interspersed by tall, slender hummocks—circular tufts of grass raised on shin-high towers of mud. I leapt awkwardly between the trees and from one hummock tower to the next, a labor in those waders that left me down in the mud more often than not.

"You won't mind, " the voice began, "if I tell you . . ."

"Stop," I said. "It's embarrassing enough without you chiming in."

". . . that you look like an old bullfrog . . ."

"Stop!"

". . . acting out his . . ."

"*Enough!* Okay?"

". . . ballerina dreams."

"*Are you done?*" I asked.

"Are *you* done?" he asked back. "You could use a break."

He was right, but there was no place to take it right then. I still had some hard work to do through the hummocks. But eventually I found shade in a sidehill grove of white pine—a grove I remembered from my winter exploration months before. Exhausted, I lay back and flexed my right ankle, feeling what I now sensed to be a sprain.

Though I didn't know for sure where I was in the swamp, I was pretty certain that I'd made it all the way around to its northwest corner and was now sitting under white pines on an east-facing slope. That meant that if I headed straight downhill from the pines and kept a steady easterly course, I'd come out close to the jewelweed, having made a complete circumference of the place. It had turned out to be pretty easy; I hadn't even needed my compass for help.

Once on the move again, I noticed in this section of swamp much more evidence of human influence: a couple of surveyor's corner pins, remnants of stone walls and barbed wire fences, and spread along a knollside to the left of me an old farm dump with its requisite scraps of machinery. Below the dump were two spring-fed pools browning with the rot of leaves.

A little further on I looked down and stopped. Something was wrong. I looked back at the sun, then turned around and stared at the ground again. It was my shadow. My shadow was wrong: I was walking right into it. For a minute I was completely stumped. It was midday, I was facing east, and my shadow was jutting out in front of me. That wasn't right. So I pulled out my compass, which confirmed the truth: I was heading north, not east.

"Oh, boy," the voice said.

"Don't even start," I said.

Practically, it was no big deal, no question of survival. I knew I could walk a straight line in any direction and make my way out of the swamp, into some place I'd recognize. But in a prideful way it was a disgrace: 90 degrees! I'd never been that turned around before, or at least I'd never been aware of it. It was like setting off from here for London and ending up in the Arctic Circle.

Keeping the same course, one I now knew to be *northerly*, I moved ahead unsteadily, hoping for a glimpse of the terrain outside the swamp to clue me in to where I was. After staggering through a stretch of reeds, I eventually came to the bank of a canal; its surface was green with algae. It was, I could tell by looking through the bordering trees, the other end of the one that had ambushed me the day before. When I stepped tentatively in to test its depth, there was a cool sensation on my left sock. I lifted up my foot to find some holes ripped in the boot. How'd I get those?

Okay, okay, I surrender, I thought. I know when to quit.

I took the fastest route out, jumping the canal where it narrowed, then limping through a last line of hemlocks and into the field north of the swamp. The car was well off to the east, almost the full width of the swamp away. When I reached it, I dropped to the ground and sat back against one of the wheels to recover.

The voice started in. "Mind if we recap the last three hours of your afternoon?" he asked.

"Nah, we don't have to," I said. "Really."

"But I bet it's somethin' Ken Bannister would do."

"Shut up," I said, trying not to crack a smile.

"Good. Okay. So. You sprained an ankle."

"Yep."

"Ripped your waders."

"Check."

"Scared all the wildlife outa the swamp."

"Now, that's an exaggeration."

"And skewed off course by ninety degrees."

"Shit. I didn't really do that, did I?"

"Afraid so. Only one success, as far as I can tell."

"What's that?" I asked.

"You came out alive."

One night two years earlier a summer storm ripped a phoebe's nest from the crossbeam under the woodshed roof. The next morning we found the nest overturned on the woodpile, and five newly hatched babies scattered like balls of gray paper on the ground. Three—one being explored by ants—were still alive and writhing, if so barely perceptible a movement can be called that, so I lifted them, rearranged them in the nest and placed the nest back on the beam.

I've never seen a form of life as fragile as those babies. No longer than my thumb, each had a tufted gray back and head and a tiny white beak and orange oval lids for eyes and beige bellyskin wrinkled like cellophane. When lifted, they drooped like a small canvas bag of children's blocks, encased but still in pieces.

The wind blew the nest down again that night, and in the morning we found only two of the babies, the third probably having slipped into the crevice between two wood stacks. Two blowdowns had been enough for them; they were dead. Carry and I buried them out back with the two others, in our animal cemetery beneath the pear tree.

The whole episode troubled me. Days afterwards I was still aching, and not just because the baby birds had died, though that was bad enough. I finally realized that the ache was for their limpness, and fragility, for their unfixable helplessness. And it twisted me with fear, the fear of inescapable doom, death around the corner—not just for myself, but for Robin, my life partner, and Carry, who was five, and Jacob, then not quite a year old, all of whom I kept envisioning swept, broken and scattered by a murderous wind none of us would see coming.

4

His strategy has worked perfectly: having drawn Robin's attention, he's made her both exasperated and tentative. Now, he can finish her off. After the boys are asleep, he comes back downstairs to the kitchen, but to his disappointment he finds her almost finished already. She's leaning back against the kitchen counter, crying softly.

"I'm afraid to tell you this," she says when she sees him. "I know you're going to get mad. But I have to say it anyway. Not just for you, but for me."

He stands across the kitchen, looking sharply at her.

"I don't know what else to do," she goes on. "What should I do? Take the kids and leave? Where would I go?" She pauses, as if there might be an answer hanging in the thick silence. "That's not my first choice. You know it's not. But I don't feel like I've been given any other one."

She looks over at him. "Ter, it's been happening again, for like the last month. You don't want to go out or do anything. You don't want to see people. You haven't really spoken to me in days. Or looked me in the eye." Her voice trails off. "You won't even touch me."

"*Don't leave!*" I try to shout, but can't. Freedom words, if they escaped, that would filter out only as a whisper, a whisper no one else might understand, but one Robin and I would both hear. A whisper that, days from now, it won't occur to me to utter.

"You're just so angry all the time. And, you know, I'm pretty sure," she says very slowly and gently, "that you're depressed again."

"*Rob, help me! Please!*" I try shouting.

"I am not!" he shouts instead. "God, I *knew* you'd say that. I was waitin' for it. You always say that. If things are fucked up between us, it's always that I'm depressed. Me. My fault. When are you gonna take responsibility for the shit you do wrong?"

"But I haven't done anything wrong!" she says loud, through tears. "And you won't ever tell me what it is you think I've done."

"Yeah, well, that's exactly the problem. I'm supposed to tell you what you've done wrong. Doesn't that strike you as a little strange? What's the matter, you can't figure it out for yourself?"

Insulted, he paces.

"Ter?" she calls to him as if he's at the far end of the house.

"What!" he yells back loudly.

"Is this about me being late tonight?"

He blurts out one sharp laugh. "Oh, yeah, right. Like I'm gonna get pissed off about somethin' like *that*."

"Then what have I done to make you so mad?"

"*What?*" he yells, louder.

"It scares me when you're this mad."

"*What!*" Louder still.

"What am I supposed to do?" she says. "It feels like I'm dying inside when you're like this."

"Dyin' inside!" he shouts, his voice exploding. "It just gets more and more dramatic with you, doesn't it?" Then he leans forward and stares right into her eyes. "How should I know what you're supposed to do. What do I look like, a fuckin' mind reader? Why don't you figure it out for yourself. This is *your* problem anyway, not mine!"

The room vibrates for a while from his anger; then, slowly, the remnants of it dissolve into the air.

And then it's quiet.

She looks at him quizzically, as if she's just run up behind him, sure of who he is, only to have him turn and reveal someone else's face. Someone she doesn't recognize. And it's then—when she realizes she's been trading yells with someone she doesn't know about something she doesn't understand—that she surrenders, backing away slowly, one soft step and then another, always facing him, her cheeks striped with tears. When she reaches the bedroom doorway, she slips inside and slowly closes the door.

Through the wall he can hear her crying, and for the first time all night he doesn't feel so angry.

Through the wall I can hear her crying and wonder how on earth I can do this to her.

Game over. He wins. He always does.

And so this is the way it will be: distance, anger, silence, sorrow. Weeks of it, maybe months. A reign of darkness and tears. Until something shifts, sometime in the future: maybe it will be Robin's voice pushing through his anger, again and again, until it finally reaches me, her words like a dawn breeze ushering the dark away. Or maybe it'll be the voice himself who relents, when the weight of his own anger gets too heavy to support and finally falls in on him like an old barn roof collapsing. Either way, on that unknowable day, I imagine myself looking up from the floor of my hostage room at the sudden shaft of light, then getting to my feet and climbing out through the hole that's been made, breathing in as if I've just been delivered.

It took me two full days to rest my ankle and pride. But on the third day I was back at the swamp, parked in the same spot and pulling through jewelweed again. It was a perfect day for redemption—sun, heavy breeze, temperature in the mid-60s.

"Well, there's good news," the voice said.

"Good," I said. "I'm in a good news kind of mood. What is it?"

"You got rid of those hideous bullfrog waders."

I looked down at my knee-high mudboots. "I know. I'm streamlined today. Feeling spry."

"Spry," he said, mulling the idea. "Okay. Whatever."

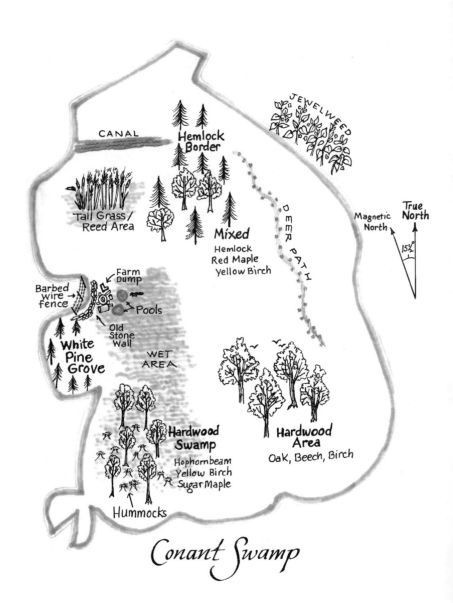

CANAL

Hemlock
Border

JEWELWEED

Tall Grass/
Reed Area

Mixed

Hemlock
Red Maple
Yellow Birch

DEER PATH

Magnetic
North

True
North

15½°

Farm
Dump

Barbed
wire →
fence

→ Pools

Old
Stone
Wall

White
Pine
Grove

WET
AREA

Hardwood
Swamp

Hophornbeam
Yellow Birch
Sugar Maple

Hardwood
Area

Oak, Beech, Birch

Hummocks

Conant Swamp

"Oh, c'mon. Lighten up. You can't be grumpy on a day like this! Look at it: cool, dry, sunny. Some autumn in the air. It's my favorite kind of day. And I'm ready to go."

While recuperating during the two previous days, I'd done everything I could to prepare for today's walk. First, I started the map: I copied the swamp's general outline from the USGS map on my office wall and then marked the features I felt sure of—jewelweed, deer path, mixed hardwoods, hummock swamp, white pine sidehill, stone wall and dump, sedge, canal, hemlock stand—giving each a different design.

Then, I determined the route I'd take today. From the minute I'd left the swamp three days before, I knew I'd have to revisit its west side, if only because I'd gotten myself so turned around there. And when I looked at my initial markings on my map, I noticed something that piqued my interest even more about that side: it appeared to be the only half of the swamp that was actually wet. That concentration of wetness hadn't occurred to me while I was walking around in it, but now it made sense: I'd noted in my field book how thickly overgrown and complicated the west side was, how the areas of hardwoods and conifers and tufted hummocks seemed blended, or else separated by borders I couldn't distinguish. And as I thought back on it, the east side of the swamp had seemed a little drier and a little more open, the vegetation less tangled. It must have had to do with the wetness of the ground underneath.

During the time that I was mulling over the strangeness of Conant Swamp—its being a swamp that was only half swampy—I came across a passage in Mitch and Gosselink's textbook on wetlands:

> Wetlands are often ecotones, that is, transition zones between uplands and deep water aquatic systems. . . . This transition position often leads to high biodiversity in wetlands, which 'borrow' species from both aquatic and terrestrial systems. Some wetlands, because of their connections to both upland and aquatic systems, have the distinction of being cited as among the most productive ecosystems on Earth.

So much for strangeness. After reading the passage, it suddenly seemed to me that if you wanted a textbook example of a wetland, you'd look no farther than Conant Swamp, because now not only could I see how perfectly it sat in a transitional plain between the uplands of Houghton Hill and High Peak and the deep water aquatic system of the Connecticut

River, but I could also see how it contained a combination of those two systems within itself. An ecotone, a mosaic of microclimates, a mixed landscape in a 90-acre oval.

Filled with those ideas today, I headed through the jewelweed, then turned west and walked across the north end of the swamp until I reached what I thought was its midline. I actually wasn't far from where I'd ended the walk three days ago, just inside the hemlock border on the north end. I was ready to follow the second stage of my Plan and explore the swamp's interior, focusing on the west half: I'd walk parallel courses, separated by about a hundred feet, back and forth through the swamp, between this north end and the south, until I reached the west edge. By the time I got to that edge, I hoped I'd have all the material I needed to finish filling out my map.

Course 1. I stood at first among yellow birch, red maple and hemlock —a mixed-wood area that seemed consistent throughout the northeast section of the swamp. It was dry underfoot, something I could tell only by feel and sound, since my feet were shrouded by fern floor. It must have happened often on my hikes, this sense of walking on invisible feet, but this was the first time I'd been aware enough to note it. And now that I did, I started worrying about the hazard of each new step, the pot-hole or tree limb or rock that might bring me rattling down.

Then I stopped. This was stupid. If I kept obsessing about each step, I wouldn't have attention to spare for anything else. So I forced myself to be quiet a minute and not to think so much, but to listen; when I did, I began to feel the gentleness of the place, the bowerlike aura of its voice: thick canopy overhead, soft floor underneath, spacious, shady, pro-tected. Quiet, with just a little breeze above. It was like Rob's breathing during sleep—deep and steady, but soft. I'd lie awake and listen to her some nights, those nights I'd stayed up late and had come to bed long after she'd drifted off. In the still dark, the sound of her breathing would put me at ease, just as the air was doing here today. It made it easier to walk on.

But the farther south I walked, the more the swamp changed. The air became windier, the ground wetter, and the understory thicker and more varied. By the time I reached the hardwood/hummock swamp, the mucky ground had sunk down as well. Or else the trees had risen. Either way, I found myself tottering on a web of elevated roots and branches, a foot or so above the ground; whenever I slipped off, I dropped shin-deep in mud.

Course 2. Balancing on a hummock at the south end, having taken my thirty paces west, and now ready to head back north, it was as if I'd come upon a completely different swamp. Loud. Mean. The kind of place I'd been afraid Conant Swamp might be all along. The wind was blowing hard in gusts, knocking the treetops together like bowling pins. Limbs made cracking sounds above me, and more than once I flinched and glanced up quickly to make sure I wasn't in the line of fire. Every time I did I became more and more aware of my solitude. I was alone out here.

And in that moment I was reminded of something that had happened that spring, when I was at a writer's conference in Montana. It was early afternoon of the first day, and I was taking a nap. I'd gone to sleep feeling a little lonely and homesick, missing Rob and the boys more than I'd thought I would. The nap was a restless one, anxious and unsettled, and somewhere out of the gale of jumbled dreams I heard Jacob's voice.

"Daddy!" he called. Not a call of fear or worry; more of excited recognition. But it was so loud and distinct that it startled me awake.

"Ja—," I began to say, before I recognized where I was—in a guest house in Montana, in a room with country-print wallpaper and drawn shades, wondering why on earth I'd gone so far from that voice and the others of home. It was as lonely as I'd felt in years.

At first the way back through the swamp was a tangled mess of roots, their latticework suspended over the mud floor. But then there came the added obstruction of dead trees, toppled, rotten, carpeted with thick moss. I leapt onto them as if they were saddled horses, then climbed down the other side.

Nothing felt stable then. Wind howling, limbs dropping, roots raised up, trees lying down. At one point, needing a rest, I put my hand up and leaned against a dead snag—a standing white birch whose top had snapped off, but whose bottom still looked solid. My first touch knocked it over and took me over with it.

Course 3. I was back in the mixed-wood area at the north end, well to the west of where I'd started. After the way the swamp had made me flinch on the last course, it might have been wisest to take the hint and quit the place for good. But here the hemlocks filtered the wind, leaving

the air shadowy and somber, not loud and wild. The atmosphere reminded me of how it had felt sometimes to rock Carry to sleep at night
when he was an infant. The rocking would have started with his wailing
discomfort but then quieted eventually to his puffy breaths in REM
sleep. And after a time even that would have given way to a deeper calm,
when all I could hear, as I cradled him across my chest and rocked forward and back, forward and back, was the muffled, rhythmic creak of
the rockers on the carpet. And though I would have been physically
exhausted from the day and emotionally spent from Carry's intensity,
there were nights when I just wanted to stay, safe and sound, in the deep
quiet of his room.

That's how I felt beneath the hemlocks. Here was a comfortable
place, the kind of place I didn't really want to leave. And when I recognized that feeling, recognized it against what I'd felt at the south end, I
knew there was something else going on in the swamp today, something
that maybe had always been going on. It was something more than an
east/west difference in moisture; there was a clear north/south distinc-
tion too. In character. In voice.

I started south. Halfway across the swamp I came to a fence, three
strands of barbed wire strung between trees. I stepped over them.
Abruptly afterward the land rose to a dry crown. Long silken grass, bent
by its own weight, lay matted in fanlike crescents. Young birch and
maple grew in thick stands there too, and with each step I had to sweep
a bunch of them aside. Once through them I met another line of barbed
wire, this time a single strand, about waist high. Beyond the wire the terrain dropped down off the crown, opened up, and grew wet again.

Even though I'd found barbed-wire strands in the most remote places
I'd walked in the valley—leading me to suspect that over the years every
square foot of land had been fenced at some time—it was still surprising to come upon it here. It was even more surprising to find two parallel fencelines so close together. Must have been a mistake, was all I could
think. It was as if someone had run that single line of wire to start, then
decided against its placement, and instead of taking it down, had just
left it there and strung a full three-wire fence a little further on, on a better line.

After stepping over the wire and off the crown, I walked a short way
before turning around to look back at it. From here I could see the fence

in broader view. I watched how the wire came downhill from the west, followed the crown's near edge, then angled away in the direction I'd come from. Though I couldn't see where it went, all at once I was pretty sure I knew: it followed the crown's curving border all the way around and back uphill. Both lines I'd crossed were actually part of the same fence—a fence that had enclosed the field uphill to the west as well. Down here, instead of making a mistake, someone had deliberately detoured the fence to hem in this narrow peninsula in the swamp. It looked like a lot of trouble for a little extra grazing area, but I could imagine what a gift it must have looked like to the fencer, this tongue of lush pasture in the swamp.

Heading away from the crown, guided by a small knoll running along my right, almost parallel to my course, I recognized where I was: I was walking the final stretch of my misdirected circumference from three days before. Today, though, I was going in the reverse direction. Shortly after, I came upon the old farm dump and the two sunken pools. Eventually I walked uphill into the white pine grove.

This was the third time in less than a year I'd come upon this grove, which I now guessed lay near the middle of the swamp's west edge, not up in the northwest corner. And I took my time there, enjoying in its confined shade the rare feeling of familiarity.

"Well, if nothing else," I said, "I recognize one place on this side of the swamp."

"I'll throw you a party," the voice said, unimpressed, "if you get home."

"You're still grumpy, aren't you?"

"Why shouldn't I be?" he answered, but not with the bite he usually had. There was something tentative in his tone. "You're all alone. You still don't know really where y'are. And the last time you walked off this hill, you headed east, except it was north."

"Wait, are you worried? Are you worried I'm going to get lost forever in here or something?"

"Oh, right. Like I'm going to be worried about somethin' like that."

"Well, if you are, don't be. I won't make the same mistake again."

Course 4. Now that I was at the west edge, I figured this would be the last course, and the hardest one too, since almost everything would be new. I hadn't been this far west before.

To get the proper spacing for it, I headed northwest, taking my paces diagonally uphill through the pine grove. That line took me outside of the trees and into an adjacent field filled with low scrub and goldenrod and a handful of apple trees. Having never been outside of the treeline, I had a little trouble getting my bearings in that field.

I stopped and, using the sun, did my best to orient myself again. Then, on instinct, I turned and headed downhill on a course that I thought was parallel to the others. But when I reached the treeline I found myself standing on the knoll above the old farm dump: I'd veered too far right and was about to cross course 3.

"Shit," I mumbled.

"Christ, just quit, will you?" yelled the voice. "I've been as patient as I can be, but you're slashin' at windmills with this place, Quixote. Face it, you're not good at compass readin' and you can't figure out the shape of the swamp. So give it up. It's pitiful. It's so pitiful you're makin' *me* embarrassed."

"No. I am *not* going to give it up," I said. I was almost there. Just one more course.

After climbing back uphill into the field, I reoriented myself further west, then started again, not downhill this time but across the field. It gave way to a sheltered grove of oak and ash sloping uphill to the left. Stopping there, I turned and looked back across the field toward the pine grove; then I looked uphill again at this grove. They seemed like mirror images, or twins.

"You're startin' to piss me off," the voice warned. "We both know you're lost again. How often are you gonna ignore me? This place owns you."

"Wait," I said to him. I had an idea.

"Whattaya think I've been doin' this whole time? I'm tired of waitin'! I'm not gonna wait anymore."

I pulled my homemade map out of my backpack and saw what I'd vaguely imagined: the west side of the swamp wasn't the gently curving line the east side was. It was double-humped, like a Bactrian camel. Now I could see it. The white pine grove was the south hump; and here I was, standing in the north one. And between them was the scrub field.

"Hey," I said to the voice. "I got it! I figured out how to read the shape of this side. I know the line for this last course!"

"You're tellin' me the two humps you're so excited about were on your map the whole time?"

"Well, yeah."

"And you're admittin' that you drew this map, right? You drew those two humps yourself, two days ago, right?"

"Yeah, but I only really knew the parts I'd walked through, which were mostly on the east side. I didn't even pay attention to the contour of this side. I mean, I'd never been in that field or this grove before. How was I supposed to know what they looked like?"

"So the real problem was that you didn't look at your map?"

"No, not the *real* problem," I said sharply. "Only part of the problem. And it wasn't a problem, because when I finally got here, I figured it out."

"And you want me to be excited about that?"

"Sure. Why not?"

"Why not?" he said, unbelieving.

"You're just disappointed I figured it out. You wanted me to stay lost so you could stay mad."

"Did not."

"Did too."

"Did not. There's plenty to be mad about: you're still makin' me wait, you haven't found your way outa here, and you definitely haven't found this mysterious answer you're so sure is in here," he said.

"Maybe this is it," I said.

"Oh, spare me," he said, sounding a little deflated. "Promise me you won't tell anybody that this is your great revelation. I couldn't bear the humiliation. Neither could you."

But, really, I thought I could. After all the bumbling I'd done in this place, I felt genuinely happy about doing something right—about sensing some of the myriad transitions within the swamp, and imagining, finally, the shape of this border and then having it confirmed by the map I'd drawn. And now, at least, I could walk out of here without feeling as if I'd been chased.

"Promise me you'll keep this to yourself," he said.

Robin's father died on our way down to see him. It was the summer before we moved to Cobble Hill. When we got to the hospital, he was lying alone in a room at the edge of the intensive care unit. Since Robin was afraid to actually see him like that, I went in by myself at first and closed the door behind me. The room was white and clean and quiet and so was he, a white sheet covering him to his neck, his face expression-less—no frown of disappointment; no crooked smirk from a joke he'd now been let in on. Just impassive, muscleless.

His figure didn't shake me; it was eerie and unsettling, but not fright-ening. I walked up to the gurney and spoke softly to him. I don't remem-ber the words exactly, but I said something about all the good things I wished for his spirit, a spirit I'd decided just then must exist, because it was too hard to explain otherwise how this man I'd known for years could lie right in front of me and yet be so clearly not there.

I bent down and looked closely into his face, as if the features might offer some hint of what to expect when my time came. Then I lifted the sheet and slipped my hand under it to touch his. I couldn't resist. I wanted to know what he felt like, wanted to understand the corporeality of death—the chill and stiffness of it. Whatever it was.

But it was nothing like that. This man, less than an hour dead, was still warm, his skin soft.

Eventually, Robin came in. She stood quietly apart from her father's body and said nothing. Later, she told me she'd sensed his spirit looking down from the far right corner of the ceiling. Unfortunately, I hadn't been sharp enough to have that sense; it would've been reassuring to feel his company then. But looking back, I can see that I'd actually felt some-thing similar, though not as vivid. What I wanted to tell Robin then, but didn't for years, was that I'd held her father's hand, tender even in death, and had marveled at the evidence of, if not an afterlife, an after*print* of life, a body in no hurry to cool and stiffen and forget the fire of the spirit that had warmed it.

5

After midnight on the first day of September, I walk halfway down Cob-ble Hill, stopping beside Asa Burton's house—now dark and empty, waiting for new owners. Just uphill from me, in the untended orchard where ripened apples have been dropping for days, I hear a soft rustle of leaves and then a solitary thump in the grass.

To the north the valley sleeps under a crystalline dark. Few lights are visible. Occasionally, there are little lantern blinks, nine miles away—headlights from cars southbound on the interstate, rolling downhill before disappearing behind the Palisades.

Later this morning I'll swallow my first antidepressant. It's odd to think it's come to this, given how helpful therapy has been for me over the last three years. Studies have shown that psychotherapy is effective in treating depression 80 percent of the time. I'd have to say I fall into that group, because I've learned more about the dynamics of myself and my moods than I thought was possible, hard lessons this valley bowl has been responsible for introducing me to during the last nine years. And I've even been able to use those lessons to lengthen the gaps between depressive episodes. But even so, the voice endures. That's probably the most pernicious aspect of depression: you can be getting effective treatment and still have it.

It was the muffled sound of Robin's crying that finally convinced me of all this, a voice of sadness that shook me awake to the fact that I needed more help. So while I'd tinkered with the idea of medicine before, this was the first time I'd committed myself to it.

And now, even though I'm still committed to following through, I'm ambivalent, if not outright scared. I'm afraid of a change I can't guess, afraid that the person who looks into the bathroom mirror after swallowing the pill won't be familiar to me, won't remember me. While I have no loyalty to the voice, I'm afraid that in creating someone new and improved, someone happier, this me will die. So just in case, I've stayed up to live out this last unmedicated night at length.

I was remarkably productive before midnight, writing letters to friends, and then messages in my journal to my medicated self. I told that future self what the voice was and had been. I warned him about the tricks the voice would use, the things he should never let himself be talked into. I hoped he'd never know the anger and self-hatred I knew, but if he did, I wanted him to be prepared. And even more, I wanted him to remember me.

Then I wrote a prose poem I called "The Belated Wake." Based on an experience I'd had canoeing one day, it came out in a rush, as few things ever have—one last lament, I guess, over the depressive drama in my head. As I finished, I saw that I'd actually been writing to Robin—my sweet love, my resilient partner, who refused to leave no matter how far away she was pushed. I guess I wanted to share with her a hope for calmer days ahead.

Tracing the bank's shadowline, I'm pulling hard upwind, upstream, in the stern of my canoe, three strokes to one side, three the other, to keep the nose from drifting, catching wind like a sail, spinning me back where I've been. An outboard pleasureboat cruises by, low and fast upchannel, crying gnat in my ear.

It's just then I pass the deer, splayed across the crown of a fallen basswood tree. Dun hide hooded and wind tattered, a smock pinned to the drying rack too long. Barkless branches support it, poking in under one side, out the other, ribs too long for the hide, skin not elastic enough for basswood bones. Its head swings free at the end of a stretched neck, mouth close to the water but not drinking.

I'm rocked in the sight of so pitiful a fit—a deer, cut and gutted, flung carelessly, like *that*, on a new frame for size.

I'm rocked in the breeze, a Canadian breeze, blowing south in my face through the valley.

I'm rocked up to the lip of this ripping night—my private feud, no one knows, voices raised behind solid doors—a lightless well from which I will not call to you and do not know how to love you.

Look at the way the dark dissolves, or seems to, with this rocking— the belated wake of the pleasureboat, now sliding noiselessly around the corner.

I face north again, toward Conant Swamp; it's sitting invisible out there, a mile and a half away—silent, noisy, tangled, clear, mucky, dry, scary, safe.

Though I never did find any of the things I imagined I might in there —nothing momentous I could pick up and bring with me—I did figure out my map, and in the end came across something that made the stumbling around inside it worthwhile. An artifact, an unlikely talisman. And like the setting for any good mystery, the swamp had held it until the very last.

I found it while walking the last course. From the oak and ash grove I veered downhill through a stand of hemlocks; white baneberries were growing like pearls at the base of their trunks. Off to the left there were some well-worn footpaths leading away from the swamp toward the homes on Latham Road, the first human paths I'd seen. Eventually, I found wetter ground again and slogged through the reedy area I'd walked through three days earlier. I stayed on its right edge, trying not to retrace any of my previous steps, but it was hard to navigate a new line, because I'd actually been there before.

Finally, I came to the canal, still blanketed with algae. Taking my time, I walked alongside it and found a place where a log had fallen across the water. I used it as a bridge. In the middle of the log, I stopped. In front of me, lying diagonally across the gray wood, like an ornament, was a dark, twisted string of weasel scat. I bent over to look at it more closely.

After a minute the voice said, "How long d'you need? It's just scat."

I stood up and laughed out loud.

"What is it?" he asked.

"This is it."

"What?"

"What I've been looking for."

"This?" he said.

"Only I didn't know it."

"This piece of shit? Oh, no. You're not gonna pick it up and take it home, are you?"

"That's totally bizarre. It was here the whole time, but I was so pre-occupied by finding my way and listening to your nagging and thinking how lonely it was in here that I never noticed it."

"Listen, I take it back," he said anxiously. "The double humps are fine. You can tell people about how you found the humps on the west side of the swamp. Really. It's fine. Just leave the weasel scat behind, okay? Good. Now, let's go home and lie down. You've exhausted your-self. You're delirious."

He was right, in a way. It was as if I'd been in some altered state of mind in the swamp, forgetting everything I'd learned about time and space and the swirl of forces animating any place. And so I'd missed the repeated messages in the swamp's voices—the winter tracks, the deer path, the surveyor's pins, the stone walls, the barbed wire and farm dump and footpaths. Robin's sleeping breaths, Jacob's call, Carry's deep-ening calm. But this scat had just reminded me, in a distinct shape I couldn't walk past, that no matter how scared or solitary or forgotten I'd felt in here, I'd never been alone or unaccompanied. Plenty of life had been here before me, was with me now, and would surely be here after.

The voice kept talking. "I can't believe you figured out those two humps," he said, with a nervousness I'd never heard before. "That was a brilliant piece of deduction. Now, step over the scat, so you don't get it on your boots, and let's get off this log before you slip and fall head first into the canal and bang your head on a rock and kill us. You don't want to die out here, do you?"

That was what he kept sounding, wasn't it? The death knell. He was right about that too, in a way. There was plenty of death around. Not just the fallen trees in the swamp or the phoebes at home, but our parents, our friends. It would be the same for ourselves and our children. Death was all around. There was no avoiding it. But it wasn't everything.

And when you joined it with life, death even took on a discernible shape, a broader contour. I could actually see its necessity, no matter how unreasonable it seemed. As Richard Nelson has said, "There is no life of *any* kind without death. Each life is nourished by thousands of little deaths, and then that life in turn becomes one of the many deaths that nourishes other life. That's the way it is."

"Because if you kill us," the voice chattered on, trying to distract me, "no one'll know where y'are. Sure, someone might see your car in the field. . . ."

But instead of being distracted, I was filled with a memory that hadn't come to mind in a long time. It was after Rob's father's funeral, when Rob and I were driving home. We were in separate cars because we'd traveled between Thetford and her family's house at different times earlier that week, but we stayed in tandem now on the way home, one of us always right behind the other.

On I-93 at the toll outside of Manchester, New Hampshire, we got separated in traffic somehow and lost sight of each other. I pulled off onto the shoulder about a quarter mile after the toll and waited for her, waited for her to pass by, waited and waited to see her, waited for about five minutes, until I had to admit that she must have driven by in the traffic and I'd missed her.

So I sped ahead, weaving through cars, hoping I might catch a glimpse of her Mustang, worried about where she was and what had happened to her and how worried *she* must have been, given the devastating weeks she'd just had. But there was no sign of her. Nothing for five miles, until the turn-off we normally took onto I-89.

I took that turn-off, determined to take the very next exit and turn around again and scour I-93 back down to the toll booth until I found her. At the next exit—for Bow, New Hampshire—I turned off, but before I could even find a way back onto the highway to head the other way, I saw the Mustang stopped across the street at the Mobil station. Miraculous.

As I drove up behind her, she got out of the car. Her cheeks were red, stained from tears.

I jumped out of my car and we ran to each other.

"I didn't know what happened to you," she said, sobbing into my chest, her arms wrapped tightly around my shoulders. "I looked up and you weren't there."

"I know, I know," I said. I was bear-hugging her, and crying too. "I pulled off to the side and waited, but I didn't see you."

"I was so scared," she went on, "after Dad and everything, and then I looked around for you and I couldn't find you and I thought I'd lost you. . . ."

"Sh, sh-sh-sh-sh," I answered softly. "That would never happen. You'll never lose me. . . ."

". . . and maybe Bob'll hear about your car being parked out there in the field," the voice was saying when I became aware of him again. "And maybe he'll remember you askin' him about the swamp, and maybe he won't. If he does, maybe he'll figure that you're lost and send the Fast Squad. But maybe he won't. I dunno. Even if he does . . ." All I could guess about this nervous chatter was that he was afraid of the weasel scat—as sincerely afraid of it as he'd been of anything, ever—afraid of what it had started me thinking about. Afraid that this might be the beginning of the end for him. ". . . it won't be for a couple of days and then you'll be all bloated and blue and putrid. . . ."

What the hell was going on? The voice was yammering away in my head while I stood on a log in Conant Swamp, hugging Robin in a gas station parking lot in Bow, New Hampshire. It must have had to do with this twist of scat, and the cortege of life and death it suddenly introduced—all of those deaths nourishing each life in order to keep the whole line going, forever passing by.

It wasn't a terrible system in theory. It might be a little unimaginative and deterministic—these life forms marching along, expending all of their energy to supply the life forms coming after them, who then expended all of their energy to supply the life forms after them, and so on. But in light of the only other existence I knew, a solitary one enslaved to a single, recurring voice of self-hatred . . .

". . . and they'll think you killed yourself, and Robin'll wonder, and the insurance company'll rig the autopsy to make it look like suicide so Robin and the boys won't get the life insurance money and. . . ."

. . . this parade of life, though not always comprehensible, was—like these stumbling rambles in the swamp, and my relationship with Rob, and my connection to Carry and Jacob, and the memories of sorrow and death, and the cacophony of voices, angry and longing and loving and calm—an abiding companionship that could pull me through the worst of times, and so, all in all, not a circumstance to despair.

I look up into the night sky now, a chilled ellipse, a wordless reflector of my own opaque heat. Orion, hunting summer while dragging winter in behind him, stands poised high above me, his bow as bright as I've ever seen. Out over High Peak, the Big Dipper tips up on its ladle.

And then. From the glint of the stars it falls—a whisper of late-autumn dusk, cold wet applescent, drenched grass, nectarine light, shadows stretched long and thin across the Evanses' field. And me stretched out too, reclined in a hammock of south-moving air, ebullient and laughing and free.

Chapter 9

The Air's Mercy

It came down in silence, like sunlight. Flew gently to the west before banking, talons open like mouths, wings flared behind. Flew across the water ten yards away and plucked the small chub from the surface as someone might lift a ring of keys off a counter. Then flew away, white head casting forward, feet swinging back, tail like white smoke behind, its brown wings beating, climbing the air like stairs.

I watched in unbreathing wonder the slow spectacle of the eagle's descent and rise, its auburn body bright, its neck yellowed like ivory. Yet for all the visual beauty, it was a sound that colored the minute most, a sound out of nowhere—a loud, thin hissing, like lips pursed and blowing, cold water poured on a hot pan, the wind through white pines. Like time passing. It was the sound of air rushing against the eagle's feathers as it crossed in front of me, wings fanned out to slow itself before the fish.

I'm sure it was surprise that made the sound so memorable. The eagle's approach had been noiseless, so I'd lulled myself into using only eyes at first. But then it swooped by, more sound than sight, and with nothing else to do, my ears, dilated like pupils, took in those hissing wings and vibrated with the sound for a long time. Even now the noise still riffles my sleep, and I can't help wondering, upon waking some mornings, whether what I'd just been hearing was the wind through those eagle feathers again, or a whisper from the valley, a passing wisdom I wasn't alert enough to catch.

*　　*　　*

6:18 A.M. The phone rings. I reach out and pick it up quickly, the way I usually do when I'm startled.

"Terry, it's Brian," the voice on the other end says.

"Hey," I say, trying to sound awake. "What's up?"

"Yeah, I just called the flight desk and they said the wind's moving in the right direction this morning. It's heading down toward your house. So I know it doesn't give you much time to get ready, but we can go up if you want."

"Hang on," I say.

Pushing up on my elbows, I look over at Robin, lying on her side, facing me, her head half buried in the pillow. She opens one eye, the only eye I can see. "It's Brian. He says it's a good morning to fly. Is that going to work? Can you take the kids?"

She nods.

"I'll be there as soon as I can," I say into the receiver, and then hang up.

For a second I pause. My heart's beating fast, the adrenaline rush from being startled awake and finding this long-awaited ride coming through. But amid the excitement, sadness is pooling. This isn't how it was supposed to happen. I'd always imagined Robin and I would go on the balloon trip together. I'd even planned for it, arranging to have Carry's and Jacob's favorite babysitter stay with them for the hours we'd be gone. I'd warned the babysitter that it would be in the early morning or the evening, and would depend on weather conditions. She'd said that was fine; she'd be prepared for a call on short notice.

But once the phone rang, my plan seemed naïve. It was unrealistic to think that we could call her out of a sleep, expect her to get dressed and ready for her day and then drive here from Fairlee in time for us to get to the Post Mills airport, where Brian would be waiting. That didn't even take into account the unknown length of the flight and the difficulties we, especially Robin, might have getting to work on time.

Shit. I could see that I hadn't really thought it through. My plans were geared toward an evening flight, when there'd be fewer time constraints. In the morning it just wasn't going to work. But this feels like the only chance I'm going to have, and I've already said yes.

As I get out of bed and start to pull on some clothes, Rob and I talk about our options and finally agree that everything considered, there's no good alternative at this point.

"I'm sorry, sweetie," I say. "I meant for us to go together. I always thought we would."

"It's okay," she says. "Maybe it's just as well. I've been kind of worried about the height anyway. I don't know if I'd get dizzy up there or not."

"Really?"

"Yeah. Really."

"You sure?" I ask again, so that I can listen to her voice for any disappointment she's trying to hide.

"Really. I'm serious."

"Okay," I say, half convinced. She sounds sincere, but I don't know. She hadn't mentioned the height before.

Maybe she *is* relieved, and she's been hiding her reluctance all along and I just hadn't heard it. Or maybe when it comes down to it she doesn't care about the ride as much as I do. Or maybe it's something else altogether. I can't tell. All I can do right now is trust the tone of her voice.

After brushing my teeth I walk quickly around the bed to her. "Brian said the wind's blowing this way, so with any luck we'll fly by. If it works out, maybe we'll even land here." I lean down, press my face into her hair and kiss her head. "I'll be looking for you guys."

"We'll look for you, too," she says. "Be safe." Then she turns over to catch another hour of sleep.

In the kitchen, waiting for the water to boil for coffee, I look through the window at the thermometer: 29 degrees. Too cold for a morning in early May.

But it's beautiful. To the north the valley lies in divided light—the Vermont piedmont glowing saffron to the west, a single star gleaming from the hillside where a house window reflects the sun, while the east side of the valley still huddles in the blue shadow of dawn, waiting for light to reach it and warm it up. Arcing over both halves is the same cloudless sky. It's going to be a gorgeous day.

It's funny to catch myself at times like these, appraising the various expressions of the valley's face and letting them influence what I think and how I behave. After ten years here, it's probably not so unusual a response. But it still surprises me; in the span of my life this attention to a landscape's cues is such a recent development, so unforeseen. And because I can't seem to get over it, I sometimes wonder whether it's *the* lesson I've been meant to learn, the one the valley has been pointing me

toward the whole time: that a landscape can, through its particular spa-
tial arrangement—the line of its hills and waterways, the pattern of
woods, the shape of sky—imprint itself on a person so deeply that the
person gradually changes, begins to take on the land's contours and
rhythms and believe in them, begins to form what psychologist Laura
Sewall calls an "ecological self."

Looking back now, I'm sure I could make that seem true, could view
the last ten years through a lens of retrospective wisdom, fit them with
this newly formed landscape philosophy, and make this morning in my
life seem a predictable, even inevitable, step in the process. But that
would be too easy, and not faithful to the truth. The truth is, I didn't
know where I was going when I started, and now that I'm here, about to
take this balloon ride that will give me an overview of the valley—a
view I've been looking forward to for years—I'm not sure how it happened. There's no map to point to. All I see when I turn around are these
hard ten years, harrowing in many ways: there I am grappling with the
voice, struggling through the swamp of my depression; there are Rob
and I bobbing through the whitewater of our marriage, trying to keep
our heads up and our hold on each other amid a current bent on separating us; there are Carry and Jacob—blossoms of Rob's and my desire,
answers to our love—now grown into flesh and blood with their own
desires, their own currents to navigate, their own valleys to set down in.
All I see now when I look out the window is a string of three thousand
days spent in the company of this place, wandering from spot to spot,
from ridges to woods to pastures, across ponds and rivers, through
swamps, learning the sights and sounds and smells, the feel of the landscape, its natural and human history, and learning about the rutted
topography of my self all the while. I wish I could be wise about how it
all happened—maybe after this balloon ride I will be—but right now,
all I know is this: I started out believing the valley had something to say
to me, and somewhere along the way I looked up and realized I was
hearing it differently, more clearly: I could understand some of its
words.

Of the language I've come to know, the weather has been the most
influential; over the years the common patterns of sun and clouds have
etched themselves inside me. Instinctively, I look left now to see the
near future, the kind of cloud formations pulling across the ridge to the

west. I've learned to see in the play of light and dark the cues for the day ahead as well—the gauzy brightness of humidity, blurring the Palisades in the distance; the cumulus shadows on fair, breezy afternoons, dotting the hills like a dalmation's spots; the morning fog in spring and fall that sits heavy on the valley floor, then swells and engulfs our house before the sun's heat pulls it apart like cotton, revealing the best days of all.

Yet, as I've also learned over the years, I'm liable to become complacent about these patterns and cues—these particular expressions from the valley, if that's what they are—so complacent that I start depending on them without knowing it, almost presuming them like breathing and setting myself up for a shock when changes come. I remember feeling dizzy the first time the air pivoted around to lead the clouds into the valley from the east; I was jarred by the lone beam of sun that broke through a thick overcast to fix on the base of the Palisades like a spotlight; I'm still disappointed by the rare summer fog that lingers into early afternoon, only to dissolve into a dreary day.

And I was startled the night I saw a strange yellow flicker in the air. It flared on, then disappeared for a moment, then flared on in a different spot before vanishing again. Wrong shape for headlights on the interstate, too low for an airplane; I couldn't tell what it was until I stepped out onto the deck and saw a round shadow, barely visible, drifting slowly east: it was Brian, stretching an evening balloon flight into darkness, the flame from his burner easing his final descent.

Following the Ompompanoosuc valley for six miles, the drive to Post Mills is quiet. Few cars on the road, none going my direction. All of the fields to my right, and the ones to my left the sun has yet to reach, are skinned with frost. The waning moon, setting to the west, shines brightly outside my driver's side window.

Frost covers the field at the airport, too, where Brian and his girl-friend Louise are rolling out the balloon just beyond the basket, which lies on its side. After parking, I walk over and help in whatever ways I can—connecting lines and then clipping on carabiners. Once the carabiners are attached, I hear a loud buzzing, the sound of the large fan Brian uses to open the balloon. Slowly the material swells. When the balloon mouth is safely open, Brian tips the basket up a little and fires his burner into it. Like an elephant getting to its feet, the balloon rises little by little until it towers over us.

It's hard to believe this is really going to happen. I've been trying to get a ride for nine years, ever since a flock of balloons flew out of my dreams and over the house.

Nine years ago they'd seeped through my sleep as sounds—intermittent, rumbling bursts. I didn't know what they were. They sounded like the surges of a big engine, and it crossed my mind that it might be a bulldozer finally digging out the driveway to the new development across the street. Gradually the bursts grew louder and more frequent, and seemed to drift around the house. That must have sent an atavistic fear through me, because in the last moments before waking—in the shallows of sleep when your mind brings the outside world into your head and bats it around, like a cat with a wounded mouse—I had the vision of a dragon standing over me, expelling a breath.

That got me up.

When I ran downstairs to the sliding door, I was met with the miraculous sight of a dozen balloons scattered in the air above the valley. Only then did I remember the announcements I'd heard on the radio that week about the Post Mills hot-air-balloon rally.

One of the balloons, only a couple of hundred feet to the east, was climbing the pines on its way south over Cobble Hill. There were more to come, all in colorful patterns—stripes, patches, geometric shapes. One was ringed by horizontal orange-and-yellow waves; another had a vertical diamond pattern in red, yellow, and blue; a third was all green with a single yellow star.

Grabbing my binoculars, I focused on a balloon that was rising right over me. I waved.

"Good morning," someone up there said. Until then it hadn't occurred to me that I might be able to hear them.

"Quite a sight to wake up to," I said.

"I hope you like it."

"I *do* like it."

And that was true: I did like it. I liked it so much, I wondered how I might snag a ride. When I noticed the yellow-star balloon landing in the field above the Asa Burton house, I saw my chance. After hurrying over and meeting the crew, who were all from Massachusetts, I watched them pack up the balloon, then gave them a ride to the airport so that they could pick up their truck and trailer, which they then drove back to Cobble Hill to fetch the balloon and basket.

I won't lie to you. Though I enjoyed meeting the crew and learning about ballooning, my motives for chauffeuring them were blatantly ingratiating: I was looking to get a ride. So that evening, when the balloons were ready to go up again, I was at the airport, standing near the yellow-star balloon, hoping Tom, its owner, might decide to take me along in trade for the lift I'd given him that morning.

As he was filling the balloon, Tom turned to me and said, "Terry, would you like to . . ."

Yes! I was thinking to myself. I couldn't believe how well my plan had worked, how deftly I'd wangled a spot in his balloon. I imagined the most gracious response I could make to his invitation: *Why, thank you, Tom. That's so kind of you. If it's convenient, I'd very much enjoy the opportunity to . . .*

". . . chase?" he asked.

"Yes!" I said enthusiastically.

No! I thought just after I'd said it. That's not what he was supposed to say. "Ride" was the right word. "Would you like a *ride*?" What the hell was *chase* anyway?

"You sure you don't mind?" he said.

"No, I guess that'd be fine." Even though I didn't know what I was getting into, what else was I going to say? I mean, I'd been trying to use him, and he'd used me instead. I sort of felt like I deserved it—a taste of my own medicine.

"Chase," as it turned out, meant tracking the balloon with the truck and trailer, and staying in contact by CB, so that the trailer could be pulled up close to the landing site. My initial disappointment aside, it was kind of exciting to be part of the team, driving familiar roads in someone else's truck, communicating on the CB, using Tom's description of the landscape below him to imagine where in the air he was.

With the wind still blowing south, the balloon headed downvalley toward our house. I drove ahead to Cobble Hill and from our deck watched the path of the yellow star as it floated closer, gliding down across Route 113 and landing in the parking lot of the state garage. God, I wished I was in that basket.

I drove the trailer down to the parking lot, where the balloon lay spread on the ground, and helped the crew pack up. Then I rode back with them to the airport, where I picked up my car. And, apart from their sincere thanks for my help, that was it. No mention of anything more. No invitation to ride with them the following day, which was the last day of the rally.

I felt a little let down, and angry at myself for being so easily taken advantage of. But the seed had been planted: I was going to find a way onto a balloon somehow, sometime.

Just as Brian's balloon finishes inflating, another person arrives for the flight. Introduced to me as Bill Saadeh, a chemistry teacher at a nearby school, he's come to collect data for an experiment he's conducting in one of his classes, an experiment studying the effect of temperature and pressure on gases. Every five minutes during the flight he's going to need to take several readings: our altitude above sea level, the pressure and temperature of the air outside the balloon, and the temperature of the air inside the balloon. For the outside readings he's brought with him a nicely carved oak barometer/thermometer unit, one he must have had hanging on a wall in his house. He'll record the interior temperature using a thermometer that's located inside the balloon. And the altitude he'll get from Brian, who's carrying an altimeter. He'll log all of the information onto the data sheet he has with him.

I like Bill immediately; he's gregarious and friendly, and we share the same discombobulation of having been in bed forty minutes ago and in a hurry ever since. Still, I'm disappointed to see him, to see anyone else. I'd assumed I'd be Brian's only passenger this morning, and have the ride to myself.

As it turns out, Bill isn't thrilled about having to be on this flight either. He's here only as a last resort, because his other plans have fallen through. He'd originally intended to have his eleven students ride in the balloon, collect the data for the experiment, and then publish the results. That was the way he thought it would happen after he'd pitched the idea to the high school administration and gotten them so excited that they agreed to cover the cost of the flights. But liability fears made them change their mind at the last minute. So here Bill is now, scrambling to collect data by himself this morning to keep this part of his course from falling through completely.

The three of us climb into the basket, and as we stand there at this delicate stasis—the balloon poised directly above, but the basket still firmly on the ground—Bill takes his first reading:

Elapsed Time: 0 mins.   Altitude: 700 ft.

Brian fires the burner and after a moment's hesitation, we're off.

The first astonishment: even though I watch us rise, I can't feel it. So gentle is the ascent that it's insensible. I'm hesitant to mention this to Brian and Bill, thinking it will surely reveal the abject dullness of my senses, so I keep my mouth shut. But thankfully, Bill reacts. He's been logging his initial readings on the data sheet and hasn't noticed that we've begun to float. Looking up, he breathes in sharply, amazed we're above the treetops.

Post Mills stretches out beneath us in shallow sunlight, the shadows still tinged with a netting of frost. As we climb higher we can see West Fairlee to the north, and to the east, the glint of Lake Fairlee.

Right away something's wrong. We're drifting northeast, opposite the direction Brian said we'd be going. Again I'm hesitant to mention it, embarrassed I might come out with something whiny like, "But you told me the wind would take us toward *my* house"; again Bill bails me out by noticing it himself. Brian agrees, a little curious about the drift himself. This leads to my first lesson about air currents: they don't move in the same direction at all altitudes. Brian says that although this lower air is taking us northeast, we should change direction and catch the higher flow of air once we're above 1,330 feet, the height of the ridge to our north.

Time: 5 mins.   Altitude: 1340 ft.

That's exactly what happens. Having gone a good distance northeast from the airport, we now begin to float slowly southeast, heading toward Lake Fairlee and the Connecticut valley.

So how is it I lived for thirty-seven years without learning that air currents move in different directions at different altitudes? And how could I have gone these last ten years without even the least suspicion of it, without guessing that this valley's swirling mosaic extends above the horizon line, that what's true for the stacked and folded land and the ebbing, flowing, eddying water must also be true for the air? It's not like this is some small detail either; it's huge.

I'll have to save that interrogation for later. There's no time to take it up now: the wind's blowing and the balloon's moving and I'm suddenly distracted by all of the sounds that are carrying up here. I'm not sure why, but I've imagined the ride as a pantomime, the land a silent movie below. Yet as we float across the shore of the lake, I can hear a dog barking as clearly as if it were next door. When I ask Brian about this, he tells us that noises don't really fade away until you climb above four thousand feet, something we won't get to experience because he won't be taking us that high this morning.

We float out over Lake Fairlee, which shows its contour clearly—it looks like a slender left hand lying open, palm up, its fingers together, its thumb extended. Directly below us the water is a dark clay color; to the east it's bright yellow, reflecting the sun. The breeze folds it all into rows of narrow ribs.

As we pass over the southeastern shore of the lake, Brian fires the burner, and we begin climbing.

Time: 20 mins.   Altitude: 2110 ft.

We sail over Ely Mountain. With the burner off, the flight is brilliantly quiet now; there isn't a single sound below that might carry up to us. On the top of the ridge the landscape looks as if it's still in the hold of late winter, probably a month behind the rest of the valley. The trees have yet to leaf, so even though the hill is thickly wooded, it looks thin, almost barren, the pale ground easily visible between trunks.

This is as close as I've been to the top of Ely Mountain in years, and I'm flooded with memories of my few days there—the circumference and diameters I walked, the knobs and ravines, the mysterious seat I came across, the deer hind leg I lifted out of the leaves and stood on its hoof, amazed to see, as I unbent and stretched it, that it reached up to my chest.

From up here those events look to me like miniature tracks, prints left in snow by a scurrying chipmunk. Except they're not a chipmunk's tracks; they're mine. And they're warm. I can feel them. It's as if, like the heat of the person no longer lying on the bed in the book *The Secret House*, some residue of my experience is lingering on Ely Mountain, rising like a column of charged air on this chilly morning, rising up to us, where we, in the balloon, happen to float through it.

And as I move through this sensation of my past, I hear philosopher Henri Bergson's words: "[The past] follows us at every instant; all that we have felt, thought and willed from our earliest infancy is there, leaning over the present which is about to join it." This morning I've felt it happen; I've felt them join.

Beyond the crest of Ely Mountain we gain our first full view of the valley, which opens slowly before us. I almost expect some fanfare to introduce the unveiling of this work, this water-drilled bowl of earth that has held the last ten years of my life. This is the view I've been waiting for all along, and to be honest I'm expecting some momentous vision, a different perspective on the valley that will change the way I see for good. Like many of my expectations, it's one I've drawn from the words of other writers.

From Isak Dinesen, who wrote: "Every time that I have gone up in an aeroplane and looking down have realized that I was free from the earth, I have had the consciousness of a great new discovery. 'I see,' I have thought, 'This was the idea. And now I understand everything.'"

From N. Scott Momaday: "Once among the mountains I flew alone in a light plane," he wrote. "I had the sense that I was entering into a great maze of the earth, in which the force of gravity was severe, broken, and erratic. The mountains exerted a crucial, irresistible force upon the air, and I was caught up in an element that I could not have imagined. . . . Never before—and not since—have I known such a feeling of buoyancy in my mind and body. It was as if the earth had let go of me and I had succeeded to some perfect equilibrium in and of the sky."

From Frank White, who coined the term "The Overview Effect" to describe the awed response he imagined people would have after seeing the earth from space: "They will be able to see how everything is related, that what appears to be 'the world' to people on Earth is merely a small planet in space, and what appears to be 'the present' is merely a limited viewpoint to one looking from a higher level."

From Apollo astronaut Rusty Schweickart, who verified what White guessed: "The whole process of what you identify with begins to shift," he said, describing the feeling of orbiting the Earth. "When you go around the Earth in an hour and a half, you begin to recognize that your identity is with that whole thing. That makes a change. You look down there and you can't imagine how many borders and boundaries you cross, again and again and again, and you don't even see them. . . . From where you see it, the thing is a whole, and it's so beautiful."

Unfortunately, from the basket my first glimpse of the valley doesn't look much different than it has from anywhere on the piedmont ridge below me. I see the bottomland and the river and the interstate. I see the New Hampshire piedmont ahead, Mount Cube and Smart's Mountain in their unchanging spots. I see the Palisades to my left, and in the distance Cobble Hill to my right. In some form or other, I've seen all of this before.

But I remind myself not to dismiss the view entirely. I know there are things to see, if only I stay patient. So I focus more closely on these familiar sights, trying to pick out things that are different or unusual. I look up at Mount Cube and Smart's Mountain again; though they're in their usual positions, I now notice that they're not their usual colorful selves this morning. They're just shadows, two broad-humped silhouettes backlit by the sun. The flat darkness makes them seem bigger than usual, more ominous.

To the north I catch a glimpse of Lake Morey's blue water beyond the Palisades. I haven't seen that from any of the ridgetop peaks.

And below, I now see the intricate designs cut into the bottomland. The contrast first—the emerald color of pastures set off against the coffee brown of the newly turned cornfields. Then I notice the way one of the cornfields is ribbed like corduroy, the straight plowlines broken only by a diagonal row of circles where the tractor has pivoted and turned around. Together, all of the fields form a geometric collage—different-sized rectangles laid out in rows, the rows divided by narrow treelines, and the whole pattern neatly rimmed to the west by the curving, black line of the river.

Upstream on the river itself, the most remarkable sight yet: the Orford-Fairlee bridge, whose steel-arch structure is visible even from Cobble Hill. This morning the bridge's crescent shape, striped by evenly spaced vertical struts—not unlike a split wagon wheel—casts a perfect, undistorted reflection on the water. I'm stunned; the reflection is so still and clear it's impossible to distinguish it from the bridge itself. In fact, with the addition of its counterimage, the structure, whose arch has always seemed beautiful on its own, now seems complete in a way it never had before. Like a butterfly that has opened its wings after holding them together on its back, the bridge with its reflection seems not doubled, but finally whole, a halved thing unfolded.

Elapsed Time: 30 mins.   Altitude: 1630 ft.

Following the contour of the ridge, we drop down into the valley and cross the interstate. The lower northbound air catches us again, changing our direction; we pick up speed in its current.

Brian could tell this was coming. When I ask him how he knew, he points toward the lumber mill in Ely, where white smoke from the smokestack stretches north like a rippling windsock. Once he points it out, I feel my perception change as dramatically as it ever has in a single moment. It's as if he's lent me his eyes: I now get a peek at his way of seeing, the priorities that stand out to him as he hovers above the landscape, the cues he uses to anticipate the future, many of them so different from mine.

The smoke trail from the smokestack. My goodness. That had never occurred to me.

So, north it is. And as we head toward Fairlee, I recognize that I've been fearing this possibility ever since we started drifting northeast from the airport in Post Mills. I didn't want to believe that the wind might actually blow us away from Cobble Hill this morning, but now I can see it's certain: we won't be going there. And all at once I'm inundated by an overwhelming sorrow.

I imagine Robin and Carry and Jacob all up and dressed, sitting at the dining room table looking through the sliding glass door at the balloon's tiny shape in the distance. I'm pretty sure that's what's happening because I've been there with them some mornings when we've watched Brian drift over the ridge. I see Carry, now a third grader, sitting at his spot at the head of the table, facing north, eating his cereal fast, his spoon circling from bowl to mouth like a paddle wheel, no stopping until he's done, while Jacob, sitting to Carry's right, eats v-e-r-y s-l-o-w-l-y, even for a three-and-a-half year old, carefully wiping each drop of spilled milk from his placemat before taking his next bite. And there's Rob, seeing the balloon in brief snatches, sometimes from the kitchen window, sometimes through the sliding glass door, as she races back and forth between the kitchen and the dining room table, trying to get all of the morning things done by herself that we usually do together: pets fed, Jacob dressed, breakfast served, lunches made, backpacks packed, teeth brushed, hair combed, shoes on, everyone and everything in the car by 8:15.

I hear the boys' voices asking where I'm going, why the balloon is drifting away when I told Mommy it was going to be coming closer. And I worry, with a grief swollen by lost time, that inside they'll register this morning's unfulfilled plan as just another disappointment from their father, whose moods can make any day unpredictable.

I don't want to be a disappointment to them anymore; I don't want to feel like one either.

With Carry almost eight and Jacob three and a half, I wonder if it's too late to rebuild a stable emotional presence for them—particularly for Carry, who's seen me at my withdrawn, distracted, irritable worst.

"You feeling alright today, Dad?" he started asking me every once in a while after I explained my depression to him. It's not a question I ever expected to hear from my seven-year-old, even from someone as good-hearted as Carry. But it's probably not a question most seven-year-olds have to worry about asking their father.

"Why?" I'd say.

"You just look a little sad," he'd answer, and most of the time he'd be right. One of the many things his inborn sensitivity has given him is a set of invisible antennae, finely tuned to other people's emotions.

At first I didn't know how to respond. Should I assure him that I was fine, thereby easing his concern but discrediting his sharp senses? Or tell him what he already suspected—that even on medication I have gloomy days now and again, days when the voice gets louder and more threatening? I made a mistake the first time: I was honest. I told him that sometimes I still did feel sad, though it had nothing to do with him and certainly wasn't his fault. Still, that was too much for him to bear: he drifted through the rest of the day worried and distracted. I watched him. I could tell.

And seeing him like that was too much for me to bear. How could I have done that—given him an up-to-the-minute reading of my depression and expected him to shoulder it? It was a hard enough load for me to haul; for him it must have been excruciating. So I promised myself the next time he asked, I'd say something like, "Actually, I'm feeling okay today, sweet guy. I must've just been daydreaming or something. But thanks for asking. I appreciate it. And I love you. You know that, right?"

That's not to say I haven't improved since taking that first antidepressant, because I have. Much of the time I've felt dramatically better. But the progress hasn't been perfect, or smooth. At first the results were remarkable; the voice disappeared for several months and I felt lighter and happier than I ever had. But after that, I sensed myself slipping, so I made the mistake of changing medication. It failed miserably: I dropped immediately into angry darkness, and it was weeks before I recognized where I was and how far I'd fallen. Another change of medication brought restored calm for about half a year and so much confidence in my stable mood that I wanted to try living without any chemical help. But without it I slipped again. So here I am, swallowing a pill a day once more, trying to regain what I've lost.

And so, like the rhythm of this balloon ride—rising and descending over the contours of the landscape, moving from one air current above to an opposite one below—my life during the last eighteen months has floated up and sunk down too, up and down, up and down, from darkness to light and back again, always moving on, but never with an advance hint of the direction.

Each time the depression returns, it gets harder for everyone, particularly Rob, and we come closer to separating. Long after the episodes we can usually step back and reassure ourselves that our struggles are normal, given the context. A study in the *Journal of Clinical Psychiatry* in 1987 reported that "recurrent major depressive disorder causes serious difficulty in the lives of close family members and friends," including "chronic marital disharmony," "chronic tension," and "fear of disease recurrence." But even so, when Rob imagines the rest of her life subjected to the voice's influence—my emotional withdrawals punctuated by unpredictable anger—she wonders whether she can endure it, and whether she'll be able resist my depression's inertia, resist being dragged down with it, a response Laura Epstein Rosen and Xavier Francisco Amador, in their book *When Someone You Love Is Depressed*, call "contagious depression."

I wonder too. She is so strong, so resilient. But only human. We both agree she shouldn't have to resist as hard as she's resisting, or endure it anymore, but when we start talking about one of us finding a separate place to live and who it would be and where and how much it would cost and where the boys would go and how we'd juggle our time with them and still keep their lives as normal as possible, and when we try to imagine what it would be like living apart from each other while still loving each other as much as we do, we fall silent, starkly aware that no alternative future looks any easier or happier than the present one.

So somehow through these episodes we manage to hang on to each other, like a pair of leaves on grafted stems, tumbling together down the face of a rogue wind whose gusts are so inevitable and tumultuous and unrelenting that there's no reasonable explanation for us still being connected. Our continued endurance of this together, in light of everything, is so unfathomable that I'll wonder occasionally whether there isn't some point to it, some reason for it all that we'll eventually learn.

In the meantime, between episodes, we cope and adapt: Robin to the threat of another onslaught; I to the sorrow and guilt I feel for hurting her and the kids over and over. And I try to reconcile myself to the vacillation I can't seem to escape: the vacillation of mood, of awareness, of knowledge, of my connection to the people closest to me. For years, ever since the time I carried water in from the well, I've recognized this inconsistency as a part of who I am and have tried to channel it to some purpose, some improvement or progress in myself. I've even tried seeing it not as abnormal behavior, but as something I have in common with the world. Henri Bergson wrote, "In reality, life is a movement, materiality is the inverse movement, and each of these two movements is simple. . . . Of these two currents the second runs counter to the first, but the first obtains, all the same, something from the second." Those words have brought me solace, even hope sometimes, since they identify vacillation as the essential pulse of life itself.

But the truth be told, I would like so much to be lifted out of this inconsistency and placed into some atmospheric stasis—to have conditions fixed, to be reliable and predictable to the people I love, to myself too. And that must be why the northward drift of the balloon this morning, away from Cobble Hill, has been so dispiriting. This one time I wanted to do what I've always dreamed of doing, what I'd hoped this balloon could help me with: I wanted to climb into the air and fly straight home.

Elapsed Time: 35 mins.   Altitude: 750 ft.

As we drop low over Ely, I ask Brian how he got started in ballooning. He says it happened by chance in 1970, when he was in New York City, finishing up his degree in art education.

"I was destined to be a teacher," he says.

While he was thinking what he might do for his final thesis, he remembered an article he'd read about ballooning, how it was coming back as a sport. And that gave him an idea: he'd build his own balloon as a creative thesis.

The point of it was more its form than its function: "Going up in it wasn't important," he tells us. He wanted to build the balloon as a sculpture.

Using a primitive burner, and a parachute cable to attach himself to the balloon, he launched his sculpture and himself from a playing field in Brooklyn. He said he bounced himself into the air a few times, the last time pretty high, before coming down and unhooking himself. He thought it had gone reasonably well. The sculpture, however, once freed from his weight, sailed off, flipped over and landed upside down on the roof of a women's dormitory, where it deflated and hung down over the building like a towel thrown over a chair.

"I asked the security guard at the front door of the dorm if I could go inside. I said I'd lost my balloon on the roof, and he was like, 'Yeah, right. Nice try. Take a hike, buddy.'"

Brian has us laughing hard as he tells us about getting his sculpture back and then about his first teaching job in Connecticut, and the accidental, unauthorized, out-of-control flight he took near Hartford, which ended in a landfill and brought the FAA after him. One unintended episode after another, and then all of a sudden there he is, ballooning for a living.

The hopscotch quality of the story makes it even funnier, especially in the context of our trio: here we are, three guys who'd never planned to be flying together, and certainly not at this particular moment. Each of us could have been doing something else this morning or been somewhere else in life, or been doing this thing in this place at some other time. The snarl of events that has brought us together in this basket at this instant is too thick to see through, too daunting to make sense of. But as we float on the crest of it, it doesn't seem anything special or unusual. Just another normal day whose events we'd have been foolish to predict.

The closer we get to the ground, the more social the flight becomes. It's as if the basket enlarges and we take on passengers for a while. A group of children, playing together in someone's front yard while waiting for the school bus, yell up to us and wave. Their brightly-colored backpacks, clustered around the base of a lamp post, look like overgrown easter eggs. On the other side of the road horses scatter in their corral, spooked by our looming presence. People come out of their houses to greet Brian, and he seems to know many of them by name. It's great to be with him and be introduced to this part of the valley, which he knows so much better than I do.

Then he fires the burner and we rise again.

Elapsed Time: 40 mins.   Altitude: 1550 ft.

We drift back across the interstate, passing the mouth of the Ely Wind Gap, heading north toward Lake Morey. Beyond the gap we float directly above the slate slope I staggered up and sat out the storm in years ago, on my way back from Oven Bird Hill; I pass through its rising memory trace just as I did the one on Ely Mountain. But there's a difference here, an added awareness; with this one's help I can finally describe what seems most different about the landscape from up here: everything seems flatter, less abrupt. The hills are bumps and ripples, the river a decorative stripe down the middle. When we pass over some steep place I've walked, I feel divided: I can remember the steepness from the experience, but I can't make my eyes see anything but the flatness.

I'm not sure what that means. I could pass it off as the idiosyncrasy of different perspectives, I suppose, and let them stand separately. But I'd rather try to keep them together, even if it's an uncomfortable fit. I'll make sure to take this vision of flatness along with me the next time I face the steepness of those slopes.

For the first time Brian talks about landing. He tells us about some small buildings he and Louise recently bought at an auction near Lake

Morey, structures he thinks will be useful at the airport if he can find a way to transport them there. As we approach the lake, he asks if we'd like to land nearby and see them. Bill says yes; he's collected enough data. I say it's fine; it sounds good to me.

But it's a lie. I'm being agreeable. What I really want is for the flight to go on and on and on. There have been so many years of anticipation for me that it seems anticlimactic to end it now, so soon after its start.

After the balloon rally nine years ago, it took me four years to call up the courage to visit Brian, introduce myself, and ask if I could earn a ride somehow. I offered to work for him, and he gave me the job of painting sealant on the airport's south annex, a two-story clapboard structure shaped like a ship's hull.

Over the next several weeks I spent parts of three days in the summer sun, balancing on an extension ladder, sealing the hull, earning a ride. And then it took four more years for the ride to happen—for Brian and me to be free at the same time, and for the weather and the wind direction to be right. He knew I wanted to float toward Cobble Hill if possible.

By my calculations, then, this 60-minute ride has had about 60,000 hours of preparation. I can't imagine why I feel any anticlimax.

Elapsed Time: 50 mins.   Altitude: 600 ft.

Sailing over the golf course just south of the lake, not far from the buildings we're going to see, something happens that shouldn't surprise us, because it's in keeping with the tenor of the entire morning. Brian realizes that he's coming in too fast and on the wrong trajectory to land where he wants to. Over the walkie-talkie he tells this to Louise, who's been chasing us in the van and trailer and is already parked next to their buildings. Then he lets her know that we're heading for Route 5, on the other side of the Palisades.

"Yeah, I should've gotten farther over," he says as we pass above Louise and the van and the trailer and the buildings.

He fires the burner.

Elapsed Time: 60 mins.   Altitude: 1110 ft.

Over Lake Morey, a lustrous green below us, we climb almost straight upward, from low fast air into the gentler current above. It bends our course to the southeast again, taking us toward the blunt top of the Palisades.

If I can't fly over our house this morning, then the Palisades isn't a bad substitute. After Cobble Hill it's the spot I'm most familiar with. Rising only five hundred feet from the valley floor, it's easy to climb in a short time; I've hiked it pretty often with Carry, the first time when he was three. From the top, especially from the cutout—a corridor cleared for powerline towers—there are long views to the south and east, including a line of sight straight down the valley toward home. I remember coming upon that perspective the first time I climbed the hill. Sitting on a rock in the cutout, training my binoculars toward home, I was amazed at how different the valley looked; it wasn't at all the place I knew from our back deck. More than any other view, the one from the Palisades pulled me off of Cobble Hill and into the landscape to see it from other angles.

Elapsed Time: 65 mins.   Altitude: 1800 ft.

Brian has taken us high for one last time this morning; we float lazily above the Palisades now, past the cutout and over the hill's beautiful cliffs, sheared off to the south and east, curved like cathedral walls.

There's probably no more dramatic spot in the valley than these cliffs. The drama doesn't have to do with privacy or quiet, as it often does with me; the interstate bends around the foot of the hill, so the cliffs resound like an echo chamber. Nor does it have to do with the scenery, really. The primary view from there takes in the paired villages of Fairlee and Orford, which are separated by the Connecticut and linked by the bridge. A beautiful view in its own way, and very different from any other in the valley, but not exactly dramatic.

No, I think what draws me to the cliffs is their combined sheerness and height. I go there and sit near the edge and feel the abrupt expansiveness of open air at arm's length. If I'm feeling brave enough, I even sit with my legs hanging over the side. It's exhilarating, but it can be unsettling, too. Especially when I look down. I feel a familiar queasiness and disorientation then; it must come from imagining myself slipping, wondering what my terror would be like in those moments of freefall. So I guess it's the intensity of the two emotions—the exhilaration of open air and the imagined terror of dropping through it—separated so delicately by only a foot of cliff's edge, that distinguishes that spot from others.

More recently I've been drawn there by peregrine falcons. Since 1992, when peregrines returned to the cliffs after a thirty-year absence, the Palisades have been occupied by a pair each year. It's now the most productive site in New England. No matter what pair nests there, they raise at least three young per year.

The peregrines mesmerize me, particularly their colors: the dark eyes, the gray hood above the beige breast, the brown-and-white-flecked belly, the yellow cere and legs and eye rings. Resting in my usual spot, I watch them climb the air in widening gyres or glide downwind below me, almost too fast to track, or perch on bare tree limbs, inspecting the traffic below, their heads flicking about in jerky motions.

What I admire most—even envy, I suppose—is something I can only imagine: their vision. John Baker writes in his book *The Peregrine*:

> The eyes of a falcon peregrine weigh approximately one ounce each; they are larger and heavier than human eyes. If our eyes were in the same proportion to our bodies as the peregrine's are to his, a twelve stone [168 lb.] man would have eyes three inches across, weighing four pounds. The whole retina of a hawk's eye records a resolution of distant objects that is twice as acute as that of the human retina. Where the lateral and binocular visions focus, there are deep-pitted foveal areas; their numerous cells record a resolution eight times as great as ours. This means that a hawk, endlessly scanning the landscape with small abrupt turns of his head, will pick up any point of movement; by focussing upon it he can immediately make it flare up into larger clearer view.

This morning, looking down at the cliffs from the basket, I think about having peregrine's eyes. If I'd had them ten years ago, I probably wouldn't have needed all this time to investigate the valley. A few trips up and back, a few across, and I'd have known the terrain more closely than I ever will with my own eyes. I wonder what alterations my brain would have to make to accommodate eyes as big as baseballs and heavy as stones, what changes would have to occur throughout my body in order to see the way a peregrine does: Would I be comfortable walking, or would I have to move faster? Would I have the same kind of conscious awareness? Would I have an unconscious? Would I construct time and space the same way? Would I ever feel depressed? And I wonder whether the trade-offs, whatever they were, would seem worth it. Right now, I can't imagine a reason for not preferring peregrine's eyes.

Drifting past the southeast edge of the Palisades, we come to a point halfway between the Vermont and New Hampshire piedmont ridges; we're positioned so that our view, as we look southwest down the valley toward Cobble Hill, seems exactly parallel to the northeast/southwest trajectory of the terrain itself. As far as I can imagine, there are only two places in this valley bowl that would offer such a symmetrical view: the air I'm drifting through now, and the air above the field just east of our house.

From this precise point and for the brief moments that we're here, the landscape stretches away in straight lines. It feels revelatory, the way it does when you're driving beside a forest that has looked for the longest time like the usual jumble of trees, but as you move along something happens: the trees seem to be organizing themselves. And they become less jumbled, more ordered, until all at once, at a specific spot, you notice they've fallen into line—neat rows separated by even spaces—and you realize that this is a nursery: the trees have been *planted*. And that single clarity in that passing moment seems to have exposed the identity of the place.

From one ridge to the other, the natural and manmade lines of the valley—the most visible features of the valley's face—lie side by side: the New Hampshire piedmont, Route 10, the river; then the railroad tracks, and Route 5, and the interstate, and finally the Vermont piedmont. But between these lines there are others visible now: with the help of the slanted morning sun and the shadows it casts, I can see the gentle terracing of the floodplain on the east side of the river, barely perceptible traces of the river's former paths. I've never seen them before.

As I widen my gaze and try to take in the whole scene instead of anything in particular, the pattern blurs. I see property lines and field boundaries cutting across the bottomland. I see houses, schools, barns, stores, and churches spread throughout. And though I know differently, though I know we've made a mess of our presence here, the valley seems to have found a comfortable serenity this morning. What I see is a culture not sawed and cleared and dug and blasted into the landscape, as I know it has been, but tucked, as if into a bed. And the ruffled topography of rock and sand and soil—made by converging plates slammed together and heated and sheared and folded, and scraped and pressed under a mile-thick sheet of ice, and cut by water—well, it just seems to have accepted our intrusion with a shrug, as if it's known a lot worse things than this.

Of course, I know that's not true. I know this valley, like the rest of the continent, would be better off today had Europeans not "discovered" it. Old-growth forests would still be here, as would the unpredictable river, whose waters raged in spring and narrowed in summer. Flocks of waterfowl would still darken the skies. Salmon and shad would fill the rivers during their spawning runs. There'd be elk, wolf, and catamount. There'd be more bears. There'd be the Abenaki, visible and in place, connected to the land's rhythms in ways we've never been.

I'm guessing that in the last quarter-millennium—from the time John Chamberlain arrived in Thetford, let's say—there have been more dramatic changes done to this land and water and air than during any similar span in the last ten thousand years. Continue the same rate of change for the next two hundred and fifty years and humans probably won't be around. Nor should we be.

Even so, this straight-lined view down the valley, the last one I'll have this morning, is telling a different story. It's making the land look resilient and human impact seem less invasive. From here the future actually looks promising.

How unexpected that idea is: a promising future. It's contrary to the things I often imagine for the earth and human existence. And for myself. So perhaps Henri Bergson is right, or if not right at least on to something: perhaps life contains simultaneous motions—an initial, enduring propulsion of spirit, and a counterpulse of matter that both resists the spirit and ushers it along. It's not a fight, not a tug-of-war like the one I've had going on inside of me for most of my life, in which two opponents try to break each other down. It's more like the two wind currents weaving the air together this morning—a slow southeast flow above a faster one heading northeast. Or like the complementary counterpressures produced by our heart as it pumps our blood: systole, diastole, systole, diastole, each pressure needing the other in order to exist. In those same ways, matter and spirit must be linked as well: spirit as the inhalation of matter; matter, the exhalation of spirit.

If that's the case, if spirit and matter are inseparable, then our spirit is bound to resonate with the land's ever-changing resilience, isn't it? Instead of trying to take material advantage of the earth, won't we even-

tually absorb our spiritual resolve from it? Somehow, sometime, won't
the land's character wash over us in torrents, slap us awake, clear our eyes
and show us that any future we have will depend on us swirling out new,
insightful patterns in our own mosaic, just as the earth has for millennia
beyond our recollecting?

The example is there. We're capable of seeing it, aren't we? Isn't that
what I've discovered from my own life, maybe the most important thing
I've discovered over and over again these last ten years—through Uncle
Asa and the mixed landscape, through my trips to the well and back,
through my unending struggle with the voice—that despite the darkness in myself and the apparent inanity of my life, despite my hard-
headed dullness at times, it's still possible to learn and change?

For this instant, near the spot where the peregrines have returned,
where salmon have trickled back, where the bear population has grown,
it does. It does seem possible for humanity to love this place for its vari-
ety and complexity, to accept it as essentially unmanageable and unpredictable. It does seem possible for us to learn about ourselves from the
contours of land and water and air, to settle into ourselves as we should
settle onto this landscape—with respect and compassion, with faith
in the unseen too—descending with wings gracefully spread, our eyes
dense but bodies light.

Elapsed Time: 80 mins.   Altitude: 390

We're coming in for a landing now, coming fast. Below us, a pale cornfield still bristles with the bases of last year's stalks. Brian has thrown the
dragline over the edge of the basket; it trails behind us like a royal-blue
tail. Louise, having watched us sail over her, runs hard through the field,
chasing the line.

After an hour I've grown more comfortable with the rhythm of the
flight. Beyond the Palisades we dropped down and drifted across Route
5, where the lower air picked us up and turned us north again. Keeping
us just above the treetops and gauging the windspeed from an American
flag flapping below us, Brian looked for a good place to land. When he
found it, he lowered us to within thirty feet of the ground, then realized
we wouldn't make it. Burner firing, we vaulted over the approaching
treeline and back across Route 5 to this cornfield. There's plenty of space
here. This is where we'll land.

So this is where I am: at the end of a morning's flight, at the end of a ten-year passage, a long way from where I started and nowhere near where I imagined I'd end up. There are dramatic differences: I'm thirty-seven now; I'm aware of my chronic depression; I'm in therapy; I take medication; I've been graced with a wife who has taken it upon herself to pull me continually from the rubble of my own internal collapses; and we have two wonderfully different sons who are becoming close friends and drawing out the best in each other.

But so much is still the same: I'm still planted on a hill inside a valley bowl, a hub inside a split wagon wheel, a high place in a low area between two raised ridgelines; I'm still sheathed by a layer of skin too thin, still being pulled back and forth, up and down, between the darkness and light of my own nature; and I'm still part of a culture that wants to lay this land bare for profit and yet somehow leave the place looking untouched, so we won't have to face what we've done. Not an easy existence to make sense of, but if it is, in fact, buffeted by the world's pulsing breath, and if that breath, like this morning's air, is moving in at least two directions at once, then I shouldn't be too surprised by what's happened.

In his book *Arctic Dreams* Barry Lopez writes:

> No culture has yet solved the dilemma each has faced with the growth of a conscious mind: how to live a moral and compassionate existence when one finds darkness not only in one's culture but within oneself. If there is a stage at which an individual life becomes truly adult, it must be when one grasps the irony in its unfolding and accepts responsibility for a life lived in the midst of such paradox.

So, maybe that's where I am now: in the midst of paradox, at the cusp of adulthood. It seems funny to imagine that here, at about the midpoint of my expected lifespan, it's time to grow up. But, hey, since I pay attention to omens, and since it's pretty clear now that I'm in this world way over my head, being steered by currents much stronger than I am, I'll take the hint.

And as I've learned from Brian and his balloon this morning, when you're at the air's mercy, there's only so much you can do. You can learn ways to see it and anticipate where it's taking you. And you can learn to maneuver your course across its currents. But that's about it. It'd be a mistake to plan anything for sure, and foolish to fight the flow altogether. You just take what you're given, and when the time comes to set down, you plant your feet and hold on.

That may seem too simple for so complicated a situation, but it's exactly what Brian has told us. "The faster we're going when we land," he said earlier in the flight, "the more the basket will want to tip and throw you out. Don't let it. We need your weight in here. So grab onto an upright."

We're below the treeline now. I watch the cornfield sling underneath us. Louise, both hands gripping the dragline, is trying to slow the drift of the basket, straining against it as if it were a wild horse. Bill is across from me, one hand clutching his barometer, his other arm wrapped around the upright nearest to him. Brian, standing relaxed against the basket, leans back a little, one arm raised, hand poised on the burner handle. Having put away my notebook and pen, and secured my binoculars around me, I now grasp the closest upright with both hands. We're all braced, ready for the land to rise up and smack us, waiting to see what happens.

In Memory of
Griffith
JUNE 24, 1988–NOVEMBER 24, 2000

For twelve years we explored this valley together, Spone.
For most of it I followed your lead, your nose.
You were an irrepressible spirit and my irreplaceable friend.
There will never be another dog like you.

3/09

Date Due